Sustainable Surfing

Whilst being an ambiguous and contested concept, sustainability has become one of the twenty-first century's most pervasive ideas, as humanity's increasing impact on the environment, as well as increasing social and economic inequalities, have local and global consequences. Surfing is a globally recognised cultural phenomenon whose unique connection with nature and rapid expansion into a multibillion pound industry offers exciting synergies for exploring various dimensions of sustainability.

This book is the first to bring together the world's foremost experts on the themes of sustainability and surfing. Drawing upon cutting edge theory and research, this book offers multidisciplinary perspectives and methodological approaches on the social, environmental and economic components of sustainable surfing. Contributions provide unique discussions that bridge the gap between theory and practice, exploring topics such as sustainable surf tourism, surf-econometrics, surf activism, surfing governance, the surfing industry, and technological advancements. Each chapter produces in-depth insights to provide foundational insights of the relationship between sustainability and surfing.

This book will appeal to multiple audiences in different disciplines and sectors. Practitioners will benefit from the insights presented in this volume, while both undergraduate and postgraduate students will find this volume an invaluable companion, including those working in geography, environmental studies, sport sciences, and leisure and tourism studies.

Gregory Borne is a Lecturer in Public Management and Policy at Plymouth University, UK. He is also the founder and Director of the Plymouth Sustainability and Surfing Research Group.

Jess Ponting is an Associate Professor and Founder and Director of the Center for Surf Research at San Diego State University, USA.

Sustainable Surfing

**Edited by Gregory Borne
and Jess Ponting**

Routledge
Taylor & Francis Group

LONDON AND NEW YORK

First published 2017
by Routledge

2 Park Square, Milton Park, Abingdon, Oxfordshire OX14 4RN
52 Vanderbilt Avenue, New York, NY 10017

Routledge is an imprint of the Taylor & Francis Group, an informa business

First issued in paperback 2018

British Library Cataloguing in Publication Data
A catalogue record for this book is available from the British Library

Library of Congress Cataloging in Publication Data
A catalog record for this book has been requested

ISBN: 978-1-138-93075-9 (hbk)
ISBN: 978-0-367-13876-9 (pbk)

Typeset in Times New Roman
by Wearset Ltd, Boldon, Tyne and Wear

Contents

List of figures   viii
List of tables   ix
Notes on contributors   x

PART I
Introduction   1

1 **Sustainability and surfing in a risk society**   3
   GREGORY BORNE

PART II
A systems approach   21

2 **Surf resource system boundaries**   23
   STEVEN ANDREW MARTIN AND DANNY O'BRIEN

PART III
Technology, industry, and sustainability   39

3 **Surfing in the technological era**   41
   LEON MACH

4 **Towards more sustainable business practices in surf
   industry clusters**   72
   ANNA GERKE

5 Surfboard making and environmental sustainability: new
   materials and regulations, subcultural norms and economic
   constraints   87
   CHRIS GIBSON AND ANDREW WARREN

PART IV
Informing policy domains   105

6 Surfing voices in coastal management: Gold Coast Surf
   Management Plan – a case study   107
   DAN WARE, NEIL LAZAROW AND ROB HALES

7 Surfers and public sphere protest: protecting surfing
   environments   125
   ROB HALES, DAN WARE AND NEIL LAZAROW

8 The non-market value of surfing and its body policy
   implications   137
   JASON SCORSE AND TRENT HODGES

PART V
Reconceptualising sustainable surf spaces   145

9 Sustaining the local: localism and sustainability   147
   LINDSAY E. USHER

10 Spot X: surfing, remote destinations and sustaining
    wilderness surfing experiences   165
    MARK ORAMS

11 Surfing: a ritual with consequences   176
    JON ANDERSON

12 Culture, meaning and sustainability in surfing   202
    NEIL LAZAROW AND REBECCA OLIVE

13 Simulating Nirvana: surf parks, surfing spaces, and
    sustainability   219
    JESS PONTING

PART VI
Conclusion 239

14 Sustainability and surfing: themes and synergies 241
GREGORY BORNE

Index 251

Figures

2.1   Surf system boundaries include physical areas and resource
      stakeholders   23
3.1   The four technological dimensions characterizing the
      technological era   51
4.1   Geographical location of the French surf industry cluster
      located in the region Aquitaine in the southwest of France   74
11.1 Uluwatu, Bali, Indonesia   181
11.2 Temples of surf   182
11.3 Do I belong here? Padang Padang   184
11.4 Surf shacks on Gili Trewangen   185
11.5 Water world resources on Gili Islands   186
11.6 Shipping in commodities   187
11.7 Bringing on water   187
11.8 Waste   188
11.9 Shipping out waste   189
11.10 Respect environment   190
11.11 How to be a responsible tourist   191
11.12 Respect culture   192
11.13 Westernising culture   193
11.14 Drug culture   194
11.15 Alcohol   195
11.16 Utopia: a place that cannot be?   198
12.1 Motivation for surfing   208
12.2 Surfers and environmental perceptions   211
12.3 What would stop you from going surfing? 1   212
12.4 What would stop you from going surfing? 2   212

Tables

1.1	Applying surfing to sustainable development and a risk society	9
1.2	Surfing and sustainability sub-categories	11
4.1	Typologies of cluster organisations in surf industry clusters	75
4.2	List of interviews	76
6.1	Strategies to manage user impact and resource base at surf locations	109
7.1	Protests against developments impacting on waves	129
7.2	Protests against developments near beaches including water quality impact	130
7.3	Protests against loss of access to surfing beaches	131
7.4	List of issues where surfers have protested against development impacting on surfing environments	132
11.1	Carbon costs of a lifestyle according to the Centre of Alternative Technology's Carbon Gym software	183
12.1	Top ten meanings or reasons for involvement in surfing activities	207
12.2	Motivation for surfing: open answers	209

Contributors

Jon Anderson is a Reader in Human Geography at Cardiff University. His research interests focus on the relations between culture, place, and identity, particularly the geographies, politics, and practices that emerge from these. His key publications include: *Understanding Cultural Geography: Places and Traces* (2010, 2015 Second Edition), *Water Worlds: Human Geographies of the Ocean* (edited with Peters, K., 2014), and *Page and Place: Ongoing Compositions of Plot* (2014).

Gregory Borne is a Lecturer in Public Management and Policy at Plymouth University. He is founder and director of the Plymouth Sustainability and Surfing Research Group. Gregory's research focuses on multiple dimensions of sustainable development at the theoretical, policy, and practical levels.

Anna Gerke is Assistant Professor at Audencia Business School and she manages the Specialised Master Programme in the Management of Sport Organisations. Her research interests are organisational theory, innovation, and strategy in sport. She has published in peer-reviewed journals like *Sport Management Review* and *European Sport Management Quarterly* and participates actively at international conferences including EURAM, NASSM, EASM, and SMAANZ. Since 2015 she is Programme Chair for the SIG Managing Sport at EURAM conference and she conducts peer reviews for various journals.

Chris Gibson is Professor of Human Geography at the University of Wollongong, Australia, and currently Editor-in-Chief of the academic journal *Australian Geographer*. His research explores cultural economies of creative, manufacturing, and rural industries, from festivals to surfboards and musical instruments. He is the author of several books including *Sound Tracks: Popular Music, Identity and Space* (2003), *Music Festivals and Regional Development in Australia* (2013), and, with Andrew Warren, *Surfing Places, Surfboard Makers* (2014).

Rob Hales is the Director of the Griffith University Centre for Sustainable Enterprise. As Director his role is to work with staff to achieve the sustainability goals of the Griffith Business School in the areas of: sustainability research, teaching and learning for sustainability, the School's sustainable operations, and engagement with sustainable enterprises and the business community. He also continues to teach in the Department of Tourism, Sport and Hotel Management. His research interests focus on social science issues in a range of contexts that include sustainable tourism, outdoor recreation, social movement studies, indigenous studies and the ethics of sustainability. He lives and surfs in Northern New South Wales, Australia.

Trent Hodges is a Surfonomics Fellow with Save the Waves Coalition and a former graduate student under Jason Scorse. Trent's research includes direct expenditure analysis of surf tourism in the World Surfing Reserves of Bahia de Todos Santos Mexico, Punta de Lobos Chile, and Huanchaco Peru. After spending two years working on the Guatemalan coast as a Peace Corps volunteer and through his research in Ensenada, Baja California Mexico, Trent realized the important role natural resource economics can play in coastal protection.

Neil Lazarow is a science-policy specialist with Australia's national scientific agency the Commonwealth Scientific and Industrial Research Organisation (CSIRO). Neil's primary research interests focus on how society makes complex trade-offs in resource constrained environments. His work extends to global development challenges such as climate, water security, sustainable cities, and food security, as well as coasts, recreation, and tourism. Neil's pioneering work in Surf Economics has helped to establish this field of research and he remains a critical thinker in this area.

Leon Mach carries a PhD in energy and environmental policy from the University of Delaware. He is the co-founder of Sea State, an organisation that runs surf-themed study-abroad programmes. He is also the associate director of the Center for Surf Research, a founding member of the International Association of Surf Academics, and a Resident Lecturer in Environmental Policy and Socioeconomic Values at the School for Field Studies in Bocas del Toro, Panama.

Steven Andrew Martin is an American academic and tenured lecturer at the Faculty of International Studies Department of Thai and ASEAN Studies, Prince of Songkla University, Phuket, Thailand. A researcher in environmental management, Steven developed the Surf Resource Sustainability Index (SRSI), a methodology and modular approach to surf site field assessment employing qualitative and quantitative metrics. His work combines personal, professional, and academic experience to highlight the value and significance of coastal surfing resources.

Danny O'Brien is an Associate Professor and Head of Program, Sport Management, in the Faculty of Health Sciences and Medicine at Bond University, Gold Coast, Australia. Danny's primary research interest is in sustainable surf tourism and its ability to contribute to community building in remote developing country contexts. His other main research interests explore event leverage, and organisational change in sport, and has been published in journals such as *Annals of Tourism Research, Journal of Sport Management, Sport Management Review,* and *Journal of Sustainable Tourism,* among others. Each of Danny's research areas shares the common sport-for-development thread and the aim to assist stakeholders in realising sustainable community-building outcomes from sport.

Rebecca Olive teaches at Southern Cross University in Australia. Her research uses feminist cultural studies approaches to explore women's experiences of recreational surfing and beach cultures, including the role of social media in constructing embodied and cultural knowledges. She has published in journals including *International Journal of Cultural Studies, Sport, Education & Society,* and *Continuum: Journal of Media and Cultural Studies,* and is the co-editor of *Women in Action Sports: Power, Identity and Experiences* (2016).

Mark Orams is the Head of School of Sport and Recreation and Associate Dean for AUT Millennium (high-performance sport) campus for the Auckland University of Technology in New Zealand. He also serves as the Founding Co-Chair of the International Coastal and Marine Tourism Society, is on the editorial boards for the journals *Tourism in Marine Environments* and *Coastal Management* and is a member of the Steering Committee for the International Congresses on Coastal and Marine Tourism. A former winner of the around-the-world yacht race, multiple world sailing champion, and Olympic coach he remains an avid surfer and waterman who is passionate about the sea and ensuring we take action to look after it.

Jess Ponting is an Associate Professor and Founder and Director of the Center for Surf Research at San Diego State University. His research interests include the sustainability of surf tourism, policy level management of surf tourism resources, surf resource governance, and the emerging surf park industry. Jess also co-founded Surf Credits (an organisation that raises funds for humanitarian and conservation NGOs in surfing destinations), STOKE Certified (a sustainability certification for surf and snow resorts), Surf Park Summit (a bi-annual conference on emerging surf park technologies and business models), and most recently the International Association of Surfing Academics.

Jason Scorse completed his PhD in Agricultural and Natural Resource Economics at UC-Berkley in 2005 with a focus on environmental economics and policy, international development, and behavioural economics. Upon graduation, he joined the faculty of the Middlebury Institute of International Studies at Monterey. He teaches courses in environmental and natural resource economics, ocean and coastal economics, behavioural economics, and sustainable development, and is Director of the Center for the Blue Economy.

Lindsay E. Usher is an Assistant Professor of Park, Recreation and Tourism Studies in the Human Movement Sciences Department at Old Dominion University in Norfolk, Virginia, USA. Her research interests include recreation conflict, surf culture and tourism, and climate change in coastal communities.

Dan Ware has a research focus on the implications of coastal management policy for distribution of political power between coastal user groups such as surfers, developers, and property owners. He is an active contributor to the development of Australian coastal management policy and practice, and is currently a Director of Surfrider Foundation Australia. He holds an applied science degree in coastal management, and an MBA. He is working on a PhD in coastal management at Griffith University.

Andrew Warren is Lecturer in Economic Geography at the University of Wollongong, Australia, and an avid surfer from the New South Wales south coast. His research interests span labour, class, and cultural and political economy. He completed a PhD on surfboard making in Australia, California and Hawai'i, and with Chris Gibson, authored *Surfing Places, Surfboard Makers* (2014), which was shortlisted for the 2015 Ka Palapala Poʻokela Awards Hawai'i Book of the Year, and won the 2015 Australian Society for Sports History Book Prize.

Part I

Introduction

1 Sustainability and surfing in a risk society

Gregory Borne

Introduction

This book will explore and expand on the relationship between sustainability and surfing. Throughout the book a number of themes and issues are evident. These include, though are not confined to, the scale and the impact of the surfing industry, the importance of the interaction between environment, society and economy, technological advancements, surfing's historical narrative, the role of surf activism and stakeholder engagement in coastal protection, issues relating to localism, overcrowding and surfing's impact on coastal environments, and the creation of artificial surfing spaces and what this means for surfing's future. This is achieved through a balance of theoretical debate, policy analysis, and practical application that builds a progressive picture of the relationship between sustainability and surfing, from the impact that surfing has in the world but also the ability of surfing to provide solutions both within the surfing zone and beyond. With that in mind this chapter will not elaborate any further on these issues or summarize the broader relationship between sustainability and surfing (see Borne and Ponting 2015; Borne 2015). Instead I jump in at the deep end and offer a theoretical frame that presents one possible perspective on sustainability and surfing.

To this end the chapter is organized in the following way achieving a number of important goals. Initially, the ambiguity of surfing is discussed highlighting surfing's connection to the natural world which perhaps forms the foundation of the relationship between sustainability and surfing. This is followed by a critical discussion of sustainable development and sustainability within the overarching context of risk, which allows a focused and succinct introduction to a vast and diversely understood concept. The focus then turns to the notion of risk itself where the seminal work of German social theorist Ulrich Beck is discussed. This inevitably elaborates on the underlying premises of the modern world emphasising in particular the often divergent perspectives of ecological modernisation and reflexive modernity. In so doing the epistemological context is established for the subsequent narrative informing not only underlying normative assumptions of sustainability but also contextualising debates in subsequent chapters. Having explored the state of modernity in a risk society I then emphasise the synergies that exist between a reflexive modernity and sustainable development suggesting that a symbiotic relationship exist between the two. This establishes a connection between theoretical speculation and empirical observation and opens up a space to explore a relationship with surfing at a very fundamental level. In order to achieve this I then provide a brief overview of existing, and where pertinent emerging areas of surfing research. This serves to provide a broad overview of surfing literature as well as furthering the overall argument of the chapter. The last part of the chapter provides an overview of the books structure. Finally, this chapter should be read in conjunction with the conclusion to this book which reengages with these discussions drawing on insights from the books contributors.

At the outset it is important to briefly explore a definition, or lack thereof, of surfing for the purposes of this chapter. It is tempting to describe surfing simply as a sport, and with an established world tour and its recent success for inclusion as an Olympic sport in the 2020 Olympic Games this is not surprising. And of course at one level surfing is a sport. However, surfing is a lifestyle activity and as such operates from within, across and beyond many categories. As Doug Booth explains, the notion of surfing as a sport '...remains a contentious subject among surfers who consider the activity a dance with a natural energy from in which the rider shares an intimate relationship with nature' (2013: 5). Anderson (2014) highlights Ford and Brown's (2006) definition that '...the core of surfing has always simply been the embodiment, raw and immediate glide or slide along the wave of energy passing through water' (2006: 149). What is crucial here in both definitions is the relationship to nature and the direct elemental contact that is the central experience of surfing. It is from this understanding that a critical relationship between surfing and sustainability extends.

This relationship between surfer and wave has been variously described as relational sensibility (Anderson 2009, 2013a, 2013b), affect (Booth 2013) or stoke (Borne and Ponting 2015). We will return to this point later in the chapter. Ultimately, it is recognised that surfing is a lose construct and that it actually has no edges or parameters from which to define it (Lazarow and Olive, this volume). This then as we will see bears a striking resemblance to the concept of sustainability and it is this acknowledgement in part, that creates an enticing and irresistible marriage between the two. Moreover, this dichotomy between sport and play will feed into core discussions in this chapter on modernity as this is '...seen as integral to the subculture's role as an agent of postmodernization within a postmodernizing mainstream' (Stranger 2011: 215). And further extending Stranger's observation on the role of the aesthetics of risk and reflexivity in surfing, this chapter will establish sustainability as a response to risk with a focus on the relationship between a risk society and sustainable development.

Sustainable development as a response to risk

Multiple reports and assessments in the past few years point to the following. The global population now stands at 7.4 billion, global greenhouse gas emissions are increasingly impacting on multiple facts of anthropogenic climate change. Biodiversity loss is continuing to accelerate, social inequality is growing and economic instability threatens social and political integrity on a global basis (UNEP 2012, 2016; UNDP 2015).

The increased use of the term sustainability, which has proliferated in the past 40 years, is a direct response to the recognition of increased risks. During this period there has been a transition of the concept from one that focuses specifically on the environment and environmental policy to one that now encapsulates the full plethora of human/environment interaction. This expansion into the three pillars of sustainable development, namely environment, society and economy, has been complemented by dimensions that include power, politics and culture. This evolution of the concept is now well documented, from the early works of Rachel Carson (1962), Aldo Leopold (1970) and Barry Commoner (1971), which can be said to have initiated the early environmentalist movement. There have also been pivotal events such as the United Nations Conference on Human Environment in 1972, the World Commission on Environment and Development (Bruntland Commission) between 1984 and 1987, the Earth Summit (1992) and the publication of Agenda 21, the World Summit on Sustainable Development through to the recent adoption of the Sustainable Development Goals (see Borne 2010, 2015; Blewitt 2015; Gupta and Vegelin 2016; Linner and Selin 2013).

In the World Commission on Environment and Development report *Our Common Future* (1987), sustainable development is defined as: 'Development that meets the needs of present populations without compromising the ability of future generations to meet their own needs' (WCED 1987: 3). Importantly, what needs to be established here is the ambiguous and contested nature of sustainable development as a concept. Within this there are two concepts. The first is the concept of needs, in particular the essential needs of the world's poor, and the second is the idea of limitations imposed by the state of technology and social organisation on the environment's ability to meet present and future needs (WCED 1987: 43). The report summarised that in order to achieve sustainable development the following would be crucial. A political system that secures effective citizen participation in decision making; an economic system that is able to generate surpluses and technical knowledge on a self-reliant and sustained basis; a social system that provides solutions for the tensions arising from disharmonious development; a production system that respects obligations to preserve the ecological base for development; a technological system that can search continuously for new solutions; an international system that fosters sustainable patterns of trade and finance; and an administrative system that is flexible and has the capacity for self-correction (WCED 1987: 65)

With the above in mind the concept has drawn considerable criticism on multiple fronts. As a concept, it has been described as an oxymoron, that no development by its very nature can be sustainable. Sustainability means so many different things to different people that ultimately it is ineffective as a concept to drive policy, implement programmes, create legislation and generally promote solutions. Perhaps the most serious accusation levelled against sustainable development is that it is a term that does nothing more than legitimise existing modes of production and consumption. This has often been termed 'green wash', where sustainable development is adopted by whatever body that might want to appear to be doing the right thing, and in different contexts these criticisms are seen to be true. But why then do we have a term like sustainable development at all and why has it now become one of the dominant concepts of the twenty-first century? I argue that there are two principle reasons for this. First, the vagueness of the concept means that it appeals to everybody. Second, whatever your opinion of the concept, there is little doubt that the direction that humanity is currently moving in is quite simply unsustainable. As a result, the idea continues to grow and embed itself in all facets of life. What is essential is to explore critically what it may mean in different contexts. Applying this understanding to sustainability in the surfing world highlights the value of exploring this relationship in multiple locales and sectors.

As well as being a response to the risks created by human interaction with the environment, sustainability has facilitated a paradigm shift, from an epistemological perspective of the way that we view and address these risks, '…as we explore ways of achieving a sustainable future, it is recognised that the problems faced by the world today and the risks that come with them, are themselves complex, uncertain non-linear crossing disciplinary boundaries, sectors and nations' (Borne 2015: 24). This has resulted in increased attention being paid to ideas of sustainability science, complex adaptive systems and idea of reflexivity. Authors in this volume either explicitly or implicitly allude to the need to adopt an approach that is complex and systemic. Most notably Martin and O'Brien, in the opening chapter to Part II, explore the idea of a resource system boundary. What follows will emphasise the idea of reflexivity and how this can inform debates on sustainability and surfing.

A risk society

Underpinning these discussions are tensions that define how modern societies operate and how this has altered over time, particularly since the Industrial Revolution. With this in mind in the following discussion I will begin to build the narrative of the relationship between sustainability and surfing though a lens of modernity as understood in a risk society. At the core of this narrative applicable to both sustainability and surfing is humanity's changing relationship to nature. As already indicated the rise of sustainable development on the global stage is a result of risk. This has led some commentators to argue that risk has now formed an organising principle within society. Seminal in this field is Ulrich Beck, who has evolved an understanding of risk through a number of key works: *Risk Society* (1992), *Global Risk Society* (1999) and finally *The Metamorphosis of the World* (2016). Beck argues that an older industrial society, whose basic principle was the distribution of goods, is being replaced by an emergent risk society, structured around the distribution of hazards. Within this analysis Beck distinguishes between three epochs of modernity. These are pre-modernity, industrial – or first – modernity and finally late – or reflexive modernity.

Broadly, the concept of modernity has been used to describe a sweeping set of social relations and processes that typify Western societies. These include science, the nation state, religion, and the family. During this time, humanity's relationship with nature is defined through domination and separation. Exponential population growth and urbanisation alter the social networks and conventional social ties within society. Politics in the modern era is defined by the nation state and a unitary political analysis where policies are created that jostle for increasing access to, and control of, the world's resources. Moreover, there is the increasing success of a capitalist market system which is a driving force of political philosophy in the modern age. Here, the acquisition of wealth is abstracted from its environmental base through the development of a monetary system.

Extending this assessment a predominant interpretation of the nature of contemporary modernistic analysis with respect to sustainable development is ecological modernisation (Hajer 1996; Mol 2000). The concept of ecological modernisation argues that the dirty and ugly industrial caterpillar will transform into an ecological butterfly (Huber 1995: 37, cited in Mol 2000). This interpretation maintains that sustainable development will be achieved within the present system of development, redefining the relationship between economy and environment in such a way that economic growth and environmental protection are seen as mutually reinforcing objectives (Blowers 1997; Mol 2000). Ecological modernisation operates on the underlying assumption that environmental crisis will spark innovation and technical development providing the necessary tools and solutions to avoid an environmental catastrophe. All that is needed is to '…fast forward from the polluting industrial society of the past to the new super industrialised era of the future' (Hannigan 1995: 185).

It is often the case that narratives of sustainable development and those of ecological modernity are conflated. Baker and Eckerberg (2008) outline four main elements to ecological modernisation. First, that there is a synergy between environmental protection and environmental growth. Second, that environmental policy is integrated throughout other areas of policy, particularly governmental. Third, that new instruments for achieving sustainability can be found, for example, voluntary agreements, pricing mechanisms, eco audit management systems. Fourth, as outlined above, ecological modernisation emphasises invention and the diffusion of new technologies.

What is argued here is that the discourses of sustainable development present a vastly more diverse and complex picture of the world. For Beck, increasing recognition of and response to the different forms of risks caused by scientific/ technological advancement and its inappropriate application by political structures is creating space for another form of modernity. According to Beck modernity has turned inward and is questioning its most central tenets, creating a stage of reflexive modernity. It is a questioning of the direction that contemporary society takes at both the global institutional scale and the local and individual level. Science and technology has lost its hegemony of knowledge formation, the relationship between established science and unconventional knowledge has become blurred, and the infiltration of the political into the scientific process disturbs the boundaries of expert and lay knowledge (Beck 1999;

Irwin 2001; Irwin and Michael 2003). In a reflexive modern world global institutions and individuals are competing for political space.

What is important to recognise in a Beckian analysis of risk is that it is the quality and not the quantity of risk that creates an epochal transition. In Beck's initial work he explores three icons of destruction. These include nuclear power, environmental degradation and genetic technology. During the course of Beck's work up to his final analysis, his risk focus has altered with a clearer focus on the role of global climate change in postmodern processes (Beck 2016). What is a central criteria for risk in a late/reflexive modern world is that all humanity is exposed, first through nuclear power and then through global climate change (1999, 2016). It is because of the global, unavoidable and inescapable nature of risk inflicted on all of humanity that a new layer of social organisation is created.

The proposition

I have previously argued that the increased use of sustainable development and sustainability within society provides an empirical litmus through which the epoch of modernity can be addressed (Borne 2010; see Redclift 1992). What is argued here is that sustainable development and the risk society present a mutually integrative storyline of humanity's influence on its environment and the complex and dynamic ways in which this relationship is reciprocated. It is suggested that both highlight particular aspects of modern developmental processes. Both explore the relationship between humanity and the environment; draw into question notions of progress, science and rationality; explore new forms of political and governance structures both above and below the nation state; open up the boundaries between the global and the local; are concerned with intergenerational equity and the incompatibility of geological and political timescapes.

With these synergies in mind I have further proposed that this leads to a symbiotic relationship between sustainable development and a risk society. From one perspective perceiving sustainable development through a risk society lens provides a level of sophistication and an overarching theoretical perspective essential for understanding the intricate and dynamic nature of sustainable development. This will ultimately lead to an informed assessment of how sustainable development is being articulated in particular contexts and the consequences this has for wider social formations. Turning that relationship around I argue that through examining sustainable development, it will be possible to directly assess some of the assertions made within the risk society thesis, particularly Beck's assertion of the emergence of a reflexive modernity.

Exploring discourses of sustainable development from within the context of the United Nations at the international level and from a localised scheme to reduce carbon emissions at the local level I have argued that '...a complex symbiotic relationship *does* exist between sustainable development and reflexive modernity' (Borne 2010: 261). Extending this logic now to surfing, what is fundamentally argued here is that examining sustainable development and sustainability within the context of surfing provides a unique opportunity to understand sustainable development and apply this to modernistic discourses. Moreover, applying this lens to the surfing world allows a sophisticated understanding of the dynamics that exist in the surfing zone but also that these are transferable to broader social, environment and economic structures. Table 1.1 identifies each symbiotic element and suggests possible surfing dynamics that can be explored.

The previous discussion has addressed a number of issues. It has introduced sustainable development and surfing through the lens of a risk society, drawing out key themes and underlying normative assumptions. In so doing it has elaborated on foundational discourses relating to sustainable development through epochs of modernity. It has established the relationship between sustainable development and a risk society and provided a link to surfing. This is not by any means a complete analysis; it is not exhaustive or definitive. It is more a sketch of the relationship that offers initial insights into the potential value of the relationship between surfing and sustainability for future scrutiny. Throughout this

Table 1.1 Applying surfing to sustainable development and a risk society

Risk society and sustainable development	Surfing
Explore the relationship between humanity and the environment	Surfing as an activity is innately connected to nature through engagement with the ocean
	Historical narrative of exploitation, domination, imperialism
	Potential reintegration of sustainability discourse and practice
Draw into question notions of progress, science and rationality	Embedded knowledge, loss of spirituality, cultural and subcultural dynamics
	Hybridity
	Wave knowledge – local knowledges
Open up the boundaries between the global and the local	Surfing's proliferation globally has impacts on multiple level scales economically, environmentally, socially, politically, culturally
	Surfing as community
	Impact of Web 2.0 on global–local boundaries
Explore new forms of political and governance structures both above and below the nation state	Multiple and expanding surf organisations that sit outside and inside the conventional political system
	The interaction of multiple organisations on a cross-sectoral basis
	Cross-sector governance dynamic
Inter-generational equity and the incompatibility of geological and political time-scapes	Protecting waves for future generations
	Surf sites

volume multiple issues and perspectives are presented that could expand and augment this framework. Equally important, there are those that could contest and contradict this assessment in order to create an informed debate. The following discussion will move to discuss more directly themes within surfing that can be said to relate to sustainability, grounding the above discussion in existing academic work.

Surfing research

Surfing as an academic pursuit is a relatively new endeavour. Each chapter in this volume explores surfing and sustainability from its own unique perspective and in so doing compiles a literature base that relates directly to surfing and sustainability as well as drawing on relevant broader literatures in order to augment and support respective discussions. With that said, because of the nascent nature of this field, clear crossovers amongst authors are visible. With this in mind, it is not the aim of this chapter to expand on this in any significant detail. It is, however, instructive to explore the range of issues that are addressed within the surfing literature and how these relate to sustainability.

Few significant attempts have been made to comprehensively explore or categorise the range of issues present within the surfing literature. Notably, Scarfe *et al.* (2009a) explore research-based literature for coastal management and the science of surfing, outlining categories and associated sub-categories that relate to this and quantifying the number of studies that relate to each. This is a very useful and comprehensive categorisation for the issues involved and with little adaptation can be modified to represent sustainability related themes.

These categories are not static or definitive – criteria in reality cross multiple boundaries and will and should evolve over time. They do, however, illustrate the breadth of issues that are present, all of which are relevant to the study of sustainability. Many of the issues illustrated in Table 1.2 are evident throughout this volume in varying degrees and in different contexts. Whilst it is not my intention here to review comprehensively the categories outlined above, or duplicate those that are explored in the following chapters, I will outline the following pertinent categories to further demonstrate a relationship between sustainability and surfing as well as risk and modernity. This predominantly focuses on surfers and wave, history, tourism, economics, sport management and sociology. In so doing areas relating to industry and coastal management are also explored.

Surfers and the wave

The discussion begins with surfers and waves, which is perhaps the most important starting point for a discussion on sustainability and surfing. This was touched upon at the beginning of this chapter, highlighting ideas of relational sensibility (Anderson 2014), affect (Booth 2013) or stoke (Borne and Ponting 2015). It has been variously argued and contested, as is the case in this volume, that the potential for surfers to act as environmental stewards and leaders

Table 1.2 Surfing and sustainability sub-categories

Category	Criteria
Surfers and the wave	Stoke and affect
	Describing waves, relating surfers to waves including skill levels, surfboard types, manoeuvres performed, surfability
Surf history	Co-evolving historical narrative
Tourism	Impacts of surf tourism on local communities
	The character or value of surf tourism
	Transportation – carbon footprint
Economics	Economic value of surfbreaks
	The blue economy
	The circular economy
Sociology	Sociological aspects of surfing including surfing culture, social protocols at surfing breaks, gender and surfing, localism
Industry	Governance, industry growth and transition, surfing equipment, technology, merchandise, marketing, clothing, surfing films and magazines and clothing
Coastal management	Coastal management theory, protecting surf breaks, recreational coastal amenities, environmental impact assessments, surfers and coastal use conflict, examples of impacts on surfing breaks
Sport management	Theories of sport management, governance and practice
Physical processes	Oceanographic and sedimentary conditions; surfing breaks including artificial breaks, hydrography, measurements
Numerical and physical modelling	Modelling of theoretical and real surfing breaks
Artificial surfing reefs (ASR) – sediment	Sediment and morphological responses to an ASR, design of ASR
dynamics, design and monitoring,	Monitoring of effects to surfing amenities, coastal stability, habitat, swimming safety
constriction	Construction techniques and monitoring
Biomechanics	Fitness, surfing techniques, sporting injuries

Source: adapted from Scarfe *et al.* (2009a).

for environmental awareness and sustainability-related issues is a natural instinct because of the direct contact with the ocean. As Whilden and Stewart maintain:

> If surfers can start to live a low carbon lifestyle and if the surf industry can develop low carbon products and practices it may be able to engineer a transformation in society itself to more rapidly engage with the CO2 problems itself and its solutions.
>
> (Whilden and Stewart 2015: 131)

The relationship between environmental contact, how we understand nature, internalise knowledge and subsequent action is complex, at both the individual and institutional levels (Macnaghten and Urry 2000; Hulme 2009; Stranger 2011). This is evident within the pages of this book, where there are different perspectives on how these variables interact. What is germane to this debate perhaps more directly is what this focus on affect means for the research agenda for sustainability and surfing. Booth suggests that the

> recent turn to affect in the social sciences and humanities among scholars which believe we should take bodies and feelings more seriously opens the door to affect as a context for surfing narratives: Indeed some authors have employed stoke as a context.
>
> (Booth 2013: 8)

Booth's comments are timely, as stoke as a context has formed the foundation for a significant assessment of narratives specifically related to sustainability and surfing (Borne and Ponting 2015). This work drew on over 40 prominent members of the surfing community on a cross-sectoral basis. This included academia, industry, not for profits, media, celebrity and government. The work highlighted the diversity and ambiguity of sustainability as a concept. It displayed a plethora of perspectives about how sustainability can be achieved from a surfing perspective but also where surfing had failed and departed from its perceived holistic, spiritual origins.

Surf history

Surf history here does not relate to the chronological emergence of surfing as a global phenomena but instead a questioning of established narratives (Endo 2015; Irwin 1973; Laderman 2014; Lawler 2011; Stranger 2011; Warshaw 2010). More recently specific lenses of surfing's history have been applied. Scott Laderman (2014) explores a political history of surfing which traverses a number of issues from the imperial roots of modern surf culture, the role of surfing in South Africa and how key surfing figures responded to the political landscape from the unique surfing narrative. Later Laderman has applied this thinking to sustainability where he critically discusses the role of the surfing industry (Laderman 2015). Focusing on surfing industry dynamics Warren and Gibson (2014) explore cultural production for the surfing industry in Australia and in this volume these authors update this assessment presenting fresh insights into surfboard manufacturing and relate this directly to sustainability.

What is evident is an innate narrative associated with surfing's history that explores the separation of surfing from the simple naiveté of the early days to the more complex structures of a modern and the postmodern world. What I suggest is that these narratives run in parallel with debates on sustainable development and the broader separation of humanity from nature through processes of industrialisation and modernisation. Drawing again on Booth as an overall historical perspective on surfing that directly feeds into the contemporary debate on sustainability the following is instructive.

> The critical questions for historians of surfing are why have surfers lost their 'sense of wonder' at the majesty of waves, and why do they no longer

respect waves, or marvel at their beauty? The immediate physical environment of surfing provides part of the answer. Today the overwhelming majority of surfers live in conurbations. Instead of escaping into nature they immerse themselves in greasy, foul-smelling waters that assault and jolt their senses. The ocean is the built environment's sewer and, like the dirty ashen skies above and the pallid concrete ribbons and blocks which abut urban beaches, it is a constant reminder of human degradation and contamination.

(Booth 1999: 52)

Booth goes further than this to emphasise that at the epistemological level there are multiple and fractured narratives that may exist at the nexus between modernity and postmodernity (2009; also see Booth 2015). This sees postmodern and reflexive modern themes of paradox, uncertainty, subjectivity and the dissolution of causality come into play and offer a further justification for viewing surfing from a risk society perspective.

Surf tourism

This departure from the naturalistic base has been predominantly accredited to the rapid rise in and commodification of surfing in the past five decades. The most visible manifestation of this, and certainly a topic that has received the most attention in the academic surfing literature, is surf tourism (Buckley 2002; Ponting *et al.* 2005; Ponting 2009; Towner 2016).

Initially, and whilst not exploring surfing tourism specifically, Buckley offers a very valuable review of the evolution and integration of sustainability into the tourism literature. Importantly it is observed that '...the key issues in sustainable tourism are defined by the fundamentals of sustainability external to the literature on sustainable tourism' (2012: 529). As such, Buckley's review applies some key components of sustainability generated externally to the sustainable tourism literature, including population, prosperity, pollution and protection. This observation applies across the themes identified here for sustainability and surfing.

With the above in mind, Ponting's work in this field has been seminal and continues to extend research within this area. This work is explored in a number of contributions, and by Ponting himself in this volume, so will not be elaborated on here. Significantly, Martin and Assenov (2012) provide a systematic review of over 5,000 pieces of literature between 1997 and 2011 related to this sub-field of surfing academia. The authors found two consistent themes within the surf tourism literature. The first is the impacts that surf tourism is having within the developing world. These studies are '...mainly directed toward capacity management in relation to social, economic and cultural interaction with host communities' (2012: 107). The second explores the central theme of urbanisation in developed countries on established surfing locations that have seen increasing numbers and echoes Booth's earlier statement. Drawing on these insights and those of Danny O'Brien, Martin and O'Brien in this volume explore the complex and integrated nature of surf systems.

Economics

Whilst intimately connected to the surf tourism literature, the economics of surfing has received increased attention. This has predominantly been articulated using the term surfonomics, which moves into areas relating environmental economics attempting to incorporate broader social and environmental issues into the economic analysis (Nelson 2015). This volume sees some exciting extensions to this work where different perspectives for incorporating value into the surfing space are discussed. Foundational and continued work by Lazarow *et al.* (2008) explores the value of recreational surfing and associated cultural impacts. Lazarow is prominent in this volume, leading on one and co-authoring two subsequent chapters from authors that have formed a cluster for surfing research at Griffith University.

Sport management

With introductory comments on the edgeless nature of surfing in the introduction to this chapter there is emerging and relevant literature in the increasingly coalesced field of sport management. There are relevant debates reacting to theory (Cunningham *et al.* 2015) and sport governance (O'Boyle and Bradbury 2014). There is an emerging body of literature that is now exploring sport and surfing more specifically within development contexts with an emphasis on community capacity (Edwards 2015; Ponting and O'Brien 2015; Able and O'Brien 2015). Also, there has been a proliferation of surfing related NGOs in the surfing space that are specifically related to community capacity but also, more broadly, sustainability-related issues. These organisations have been termed Surfing Development Organisations and relate to multiple areas of relevance to sustainability (Borne and Ponting 2015). These include social issues, inclusion, gender, and environmental protection (Britton 2015; Schumacher 2015; also see Roy 2014), market and culture transformation (Whilden and Stewart 2015), and coastal land protection (Dedina *et al.* 2015).

Sociology and an age of sustainability

The final category I want to explore is sociology, which forms a prominent perspective on surfing, particularly when applied to discussions relating to culture and subculture. Ford and Brown (2006) have drawn together social theory and surfing in an effective framework that weaves a number of traditional sociological approaches with a broad range of surf literature. This work has gone a great distance to taking a serious academic approach to the nature of surfing. Using a combination of popular surf culture, academic literature and social theory, the book discusses the contemporary social and cultural meaning of surfing.

Pertinent areas include mind and body, emotions and identity, aesthetics, style and sensory experience. Key themes include evolving perceptions of the sea and the beach, the globalization of surfing as a subculture and lifestyle and the embodiment and gendering of surfing. Lawler (2011), extrapolates the narrative of surfing as subculture or radical culture, exploring the relationship this has with capitalism, with an emphasis on the United States.

Risk, modernity and a space for sustainability

Mark Stranger's (2011) work represents arguably one of the most comprehensive, analytical and thought-provoking assessments of surfing culture, its relationship to industry as well as a discussion on the future of surfing, which is addressed though the context of risk and the aesthetics of surfing. As such, in a limited way, with what follows I want to highlight the relevance of Stranger's analysis for the relationship between sustainability and surfing and the risk society which represents a synthesis of his work and my own. For Stranger there is no doubt that surfing culture has been appropriated for popular consumption. However, what Stranger highlights is that the progression of surfing culture does not exist in the polemic of culture and subculture. Instead there was a counter-cultural movement that developed in a co-emergent way where the surfing culture industry led to a post-Fordist incarnation and not into the idea of big capitalism.

What Stranger explains is that the dominant industry players remain insiders to the foundational experience of surfing, which has '... made them bulwarks of resistance to surfing's subsumption within the dominant culture' (Stranger 2011: 190). This complex co-evolution of the big business commodification of surfing culture (and in tandem nature) present important lessons for how the surfing world may not only be informed to be more sustainable from an impact perspective but also how it can utilise this co-evolution of sub- and mainstream culture to project sustainability to a broader global community on a cross-sectoral basis. At this point in the chapter we see a convergence on multiple fronts.

First, synergies are now evident with the broader discussion on the state of modernity as understood in a risk society. Stranger explains that the surfing industry connection to its subculture is the result of an internal reflexivity which acknowledges the 'need' for subcultural maintenance for its own survival. For Stranger this is oppositional postmodernism created by the shared foundational experience of surfing that underpins the subculture. 'However, this postmodern form is in conflict with a concurrent mainstream postmodernism which favours market governance and individual distinction and through which the surfing culture industry is also linked through the arena of competition surfing' (2011: 214).

If we now insert sustainability into this analysis as understood in a risk society we can directly overlay the discussion on ecological and reflexive modernity onto oppositional postmodernism and mainstream postmodernism. In an era of sustainability the internal reflexivity once again creates a situation of survival as threats such as climate change impose directly (perceived or actual) on the foundational experience of surfers, which ultimately distorts the maintenance of a subculture. This then is the space that the relationship between sustainability and surfing now occupies. The conclusion to this book will further develop these discussions drawing on the contributions contained within this text.

Structure of the book

This book is divided into six parts including this introduction and the conclusion. In Part II Steven Martin and Danny O'Brien introduce and develop the concept of system boundaries. This is a theoretical concept in environmental science that represents intersecting, interrelated human and physical elements at a given site. This is a natural opening chapter as it explodes the rubric of what constitutes a surf site, highlighting the interrelated nature of multiple variables.

Part III introduces three chapters that focus on the relationship between, technology, industry, surfing and sustainability. The section opens with a broad-based chapter by Leon Mach that explores the technological epoch of surfing. This chapter builds a sophisticated understanding of the role of technological advance. It establishes a better understanding of the ways in which the changing technological parameters facilitate and/or detract from efforts to usher in more sustainable surfing-related practices. The chapter focuses on four technological dimensions impacting surf tourism, including the physical, climatology, Internet communication technologies and artificial surfing. Anna Gerke deals with the local development of the French surf industry in the Aquitaine region and its impacts on the local economy. Tendencies towards more sustainable business practices are outlined using practical examples from the region. Chris Gibson and Andrew Warren conclude this section by discussing environmental sustainability issues in the surfboard-making industry, and dilemmas that arise as a consequence of the industry's combination of structural economic features and subcultural origins. The authors highlight issues such as the dependence on petroleum products, harmful chemicals and poor waste management practices.

Part IV focuses on the inclusion of surf activism and non-traditional economic accounting in the policy domain. Dan Ware *et al.* highlight the inauguration of the world's first ever surf council on Australia's Gold Coast. Exploring power dynamics the authors emphasise that well educated, connected communities have the greatest success in maintaining their local environments and are generally avoided by development interests.

Rob Hales *et al.* explore the literature on surfing and public sphere resistance to development. Again focusing on Australia's Gold Coast, the authors theoretically position surfers' resistance to development as the unique feature of the 'surfing common' as the sphere of action for the pursuit of sustainability governed by public interest. The final chapter in this section by Jason Scorse and Trent Hodges compliments the previous chapters by highlighting the non-market value of surfing and its policy implications through the Hedonic Price Method.

The common theme in Part V is the reconceptualisation of surf spaces. Lindsay Usher opens the section by discussing the implications of localism for the sustainability of surfing and surf travel. This chapter outlines sustainability and surfing and explores the positive and negative impacts localism can have on achieving sustainability. Mark Orams draws into question the sustainability of surf tourism, emphasising that managing growth to ensure that a range of surfing experiences remain available, from high-use multi-sport venues to wilderness solo surfing venues, is an important challenge for the future of surfing as its popularity continues to grow worldwide. Jon Anderson looks to reframe the sustainability of a surfing utopia. The chapter introduces the place of the tube, outlining the cultural and social costs many places suffer through the influx of surfers into an area. The chapter examines the carbon footprints involved in travel and their significance for the ecological world. Anderson suggests how lessons from environmentalism, including new transitional politics of pragmatism, could be usefully adopted by surfers to better frame the surfing dream as inevitably tied into the representations and structures of a currently unsustainable world. In Part VI I extend the analysis presented in this introduction whilst summarizing author contributions.

Neil Lazarow and Rebecca Olive explore key aspects of the contemporary surfing world, including culture, meaning, place and sustainability. The chapter presents primary data and highlights the complex dynamics in relation to surfers engaging with environmental issues. In the final chapter of this section Jess Ponting discusses the relationship between surf parks, surfing spaces and sustainability.

References

Able, A. and O'Brien, D. (2015) Negotiating communities: sustainable cultural surf tourism. In G. Borne and J. Ponting (eds), *Sustainable Stoke: Transitions to Sustainability in the Surfing World*. Plymouth, UK, University of Plymouth Press: 154–165.

Anderson, A. (2009) Transient convergence and relational sensibility: beyond the modern constitution of nature. *Emotion, Space and Society* 2: 120–127.

Anderson, A. (2013a) Surfing between the local and the global: identifying spatial divisions in surfing practice. *Transactions of the Institute of British Geographers* 39: 237–249.

Anderson, A. (2013b) Cathedrals of the surf zone: regulating access to a space of spirituality. *Social and Cultural Geography* 14(8): 954–972.

Anderson, A. (2014) Exploring the space between words and meaning: understanding the relational spaces. *Emotion Space and Society* 10: 27–34.

Baker, S. and Eckerberg, K. (eds) (2008) *In Pursuit of Sustainable Development: New Governance Practices at the Sub-national Level in Europe*. New York, Routledge.

Beck, U. (1992) *Risk Society: Towards a New Modernity*. London, Sage.

Beck, U. (1999) *World Risk Society*. Cambridge, Polity Press.

Beck, U. (2016) *The Metamorphosis of the World*. Cambridge, Polity Press.

Blewit, J. (2015) *Understanding Sustainable Development*. London, Routledge.

Blowers, A. (1997) Environmental policy: ecological modernisation of the risk society: the quest for sustainable development. *Urban Studies* 30(5): 775–796.

Booth, D. (1999) Surfing: the cultural and technological determinants of a dance. *Culture, Sport, Society* (2)1: 36–55.

Booth, D. (2009) Sport history and the seeds of a postmodernist discourse. *Rethinking History* 13(2): 153–179.

Booth, D. (2013) History, culture, surfing: exploring historiographical relationships. *Journal of Sport History* 40(1): 3–20.

Booth, D. (2015) The myth of Bondi's Black Sunday. *Geographical Research* 53(4): 370–378.

Borne, G. (2010) *A Framework for Sustainable Global Development and the Effective Governance of Risk.* New York, Edwin Mellen Press.

Borne, G. (2015) Sustainable development and surfing. In G. Borne and J. Ponting (eds), *Sustainable Stoke: Transitions to Sustainability in the Surfing World.* Plymouth, UK, University of Plymouth Press: 18–27.

Borne, G. and Ponting, J. (eds) (2015) *Sustainable Stoke: Transitions to Sustainability in the Surfing World.* Plymouth, UK, University of Plymouth Press.

Britton, E. (2015) Just add surf: the power of surfing as a medium to challenge and transform gender inequalities. In G. Borne and J. Ponting (eds), *Sustainable Stoke: Transitions to Sustainability in the Surfing World.* Plymouth, UK, University of Plymouth Press: 118–127.

Buckley, R. (2002) Surf tourism and sustainable development in Indo Pacific Islands I: the industry and the islands. *Journal of Sustainable Tourism* (10): 405–424.

Buckley, R. (2012) Sustainable tourism: research and reality. *Annals of Tourism Research* 39(2): 528–546.

Carson, R. (1962) *Silent Spring.* London, Penguin Books.

Commoner, B. (1971) *The Closing Circle: Nature, Man and Technology.* New York, Alfred Knopf.

Cunningham, G., Fink, J. and Doherty (eds) (2015) *Routledge Handbook of Theory in Sport Management.* London, Routledge.

Dedina, S., Najera, E., Plopper, Z. and Garcia, C. (2015) Surfing and coastal ecosystem conservation on Baha California, Mexico. In G. Borne and J. Ponting (eds), *Sustainable Stoke: Transitions to Sustainability in the Surfing World.* Plymouth, UK, University of Plymouth Press: 166–171.

Edwards, M. (2015) The role of sport in community capacity building: an examination of sport for development research and practice. *Sport Management Review* (18): 6–19.

Endo, T. (2015) Crimes committed in the spirit of play. In G. Borne and J. Ponting (eds) *Sustainable Stoke: Transitions to Sustainability in the Surfing World.* Plymouth, UK, University of Plymouth Press: 84–88.

Ford, N. and Brown, D. (2006) *Surfing as Social Theory: Experience, Embodiment and the Narrative of the Dream Glide.* New York, Routledge.

Gupta, J. and Vegelin, C. (2016) Sustainable development goals and inclusive development. *International Environmental Agreements: Politics Law and Economics* 16(3): 433–448.

Hajer, M. (1996) Ecological modernisation and cultural politics. In S. Lash, B. Szersznski and B. Wyanne (eds) *Risk, Environment and Modernity: Towards a New Ecology.* London, Sage: 246–268.

Hannigan, J. (1995) *Environmental Sociology: A Social Constructivist Perspective.* London, Routledge.

Hulme, M. (2009) *Why We Disagree about Climate Change: Understanding Controversy, Inaction and Opportunity.* Cambridge, UK, Cambridge University Press.

Irwin, A. (2001) *Sociology and the Environment: A Critical Introduction to Society, Nature and Knowledge*. Malden, MA, Polity Press.

Irwin, A. and Michael, M. (2003) *Science, Social Theory and Public Knowledge*. Maidenhead, UK, Open University Press.

Irwin, J. (1973) Surfing: the natural history of an urban scene. *Urban Life and Culture* 2(2): 131–160.

Laderman, S. (2014) *Empire in Waves: A Political History in Surfing*. London, University of California Press.

Laderman, S. (2015) Beyond green: sustainability, freedom and labour of the surf industry. In G. Borne and J. Ponting (eds), *Sustainable Stoke: Transitions to Sustainability in the Surfing World*. Plymouth, UK, University of Plymouth Press: 80–83.

Lawler, K. (2011) *The American Surfer: Radical Culture and Capitalism*. London, Routledge.

Lazarow, N., Miller, M. and Blackwell, B. (2008) The value of recreational surfing to society. *Tourism in Martine Environments* 5(2–3): 145–158.

Leopold, A. (1970) *A Sand County Almanak*. New York, Oxford University Press.

Linner, B. and Selin, H. (2013) The United Nations Conference on Sustainable Development: forty years in the making. *Environment and Planning C: Government and Policy* 31: 971–987.

Macnaghten, P. and Urry, J. (2000) *Contested Natures*. London, Sage.

Martin, S. and Assenov, I. (2012) The genesis of a new body of sport tourism literature: a systematic review of surf tourism research (1997–2011). *Journal of Sport & Tourism* 17(4): 257–287.

Mol, A. (2000) The environmental movement in an era of ecological modernisation. *Geoforum* 31: 45–56.

Nelson, C. (2015) Surfonomics: using economic valuation to protect surfing. In G. Borne and J. Ponting (eds), *Sustainable Stoke: Transitions to Sustainability in the Surfing World*. Plymouth, UK, University of Plymouth Press: 104–109.

O'Boyle, I. and Bradbury, T. (eds) (2014) *Sport Governance: International Case Studies*. London Routledge.

Ponting, J. (2009) Projecting paradise: the surf media and the hermeneutic circle in surfing tourism. *Tourism Analysis* 14(2): 175–185.

Ponting, J. and O'Brien, D. (2015) Regulating Nirvana: sustainable surf tourism in a climate of increasing regulation. *Sport Management Review* 18(1): 99–110.

Ponting, J., McDonald, M. and Wearing, S. (2005) Deconstructing wonderland: surfing tourism in the Mentawi Islands, Indonesia. *Society and Leisure* 28: 141–161.

Redclift, M. (1992) Sustainable development: needs values, rights. *Environmental Values* 2(1): 3–20.

Roy, G. (2014) Taking emotions seriously: feeling female and becoming – surfer through UK Surf Space. *Emotion Space and Society* 12: 41–48.

Scarfe, B., Healey, T. and Rennie H. (2009a) Research based literature for coastal management and the science of surfing – a review, *Journal of Coastal Research* 25(3): 589–557.

Scarfe, B.E., Healy, T.R., Rennie, H.G. and Mead, S.T. (2009b) Sustainable management of surfing breaks: case studies and recommendations. *Journal of Coastal Research* 25(3): 684–703.

Schumacher, C. (2015) Shifting surfing: towards environmental justice. In G. Borne and J. Ponting (eds), *Sustainable Stoke: Transitions to Sustainability in the Surfing World*. Plymouth, UK, University of Plymouth Press: 90–92.

Stranger, M. (2011) *Surfing Life: Surface, Substructure and the Commodification of the Sublime*. London, Routledge.

Towner, N. (2016) Searching for the perfect wave: profiling surf tourists who visit the Mentawi Islands. *Journal of Hospitality and Tourism Management* 26: 63–71.

United Nations Development Programme (UNDP) (2015) *Human Development Report 2015: Work for Human Development*. http://report.hdr.undp.org/, accessed 1 March 2016.

United Nations Environment Programme (UNEP) (2012) *GEO 5: Environment for the Future We Want*. http://web.unep.org/geo/assessments/global-assessments/global-environment-outlook-5, accessed 13 November 2016.

United Nations Environment Programme (UNEP) (2016) *GEO 6: Healthy Planet Healthy People*. http://web.unep.org/geo/, accessed 13 November 2016.

Warren, A. and Gibson, C. (2014) *Surfing Places, Surfboard Makers*. Honolulu, HI, University of Hawai'i Press.

Warshaw, M. (2010) *The History of Surfing*. San Francisco, CA, Chronicle Books.

WCED (World Commission on Environment and Development) (1987) *Our Common Future*. Oxford, Oxford University Press.

Whilden, K. and Stewart, M. (2015) Transforming surf culture towards sustainability: a deep blue life. In G. Borne and J. Ponting (eds), *Sustainable Stoke: Transitions to Sustainability in the Surfing World*. Plymouth, UK, University of Plymouth Press: 130–139.

Part II

A systems approach

2 Surf resource system boundaries

Steven Andrew Martin and Danny O'Brien

A 'system boundary' is a theoretical concept in environmental science representing the intersecting and interrelated human and physical elements in the natural world at a given site. This chapter develops a system boundary discussion on surf sites, recognizing 'surf system boundaries' (Figure 2.1) as more than the beach and sea; they encompass numerous stakeholder interests and factors related to the scope of the 'whole' surf system as a sustainable and dynamic model. The following discussion serves to review and broaden the knowledge of surf system boundaries and provide clarity in two sets of dimensions: the physical boundaries of surf sites and the resource stakeholders.

Physical dimensions

Over the last 30 years, it has increasingly been recognized that mankind's economies and even survival are challenged by the realities of ecological and

Figure 2.1 Surf system boundaries include physical areas and resource stakeholders.

economic interdependence – and nowhere is this more true than in shared eco-systems and in 'the global commons', such as the oceans and in particular shore-lines. The UN report *Our Common Future* (United Nations, 1987) emphasizes that the oceans cover over 70 per cent of the planet's surface and provide the balance in the Earth's wheel of life: 'They play a critical role in maintaining its life-support systems, in moderating its climate, and in sustaining animals and plants, including minute, oxygen-producing phytoplankton ... they provide protein, transportation, energy, employment, recreation, and other economic, social, and cultural activities' (1987: 179).

Thus, the oceans are marked by a fundamental unity from which there is no escape, where interconnected cycles of energy, climate, marine living resources, and human activities move through coastal waters (United Nations, 1987). Coastal areas, such as beaches, along with the accompanying dunes and shore-line environments, were established after stabilization of sea level less than 7,000 years ago and are part of an interconnected single natural system (GOP, 2013). Surf sites are dynamic features of the littoral, comprised of a particular set of geographic features and phenomena that unite the physical system in such a way that waves form and break in a manner that is conducive to surfing. They include the surf zone (the area where waves break as they approach the shore) as well as the areas affected by local tides and local flora and fauna and are part of a wider natural system (GOP, 2013). The physical dimensions of sites include the sea and the waves, the beach and sand bars, the reefs and biodiversity, the adjacent terrestrial environment and a number of physical processes.

Research accounting for the wider natural surf system has only recently appeared in the literature, particularly in reports by the not-for-profit sector (Martin and Assenov, 2012). Increasingly, the physical dimensions of surf sites, including geomorphic and bathymetric features, are being recognized as baseline to the integrity of sites (Bicudo and Horta, 2009; Scarfe *et al.*, 2009; Surfrider Foundation, 2016a). Accordingly, the physical boundaries of surf sites encom-pass more than the littoral, and their integrity is linked to and dependent on adja-cent terrestrial areas and open sea. For example, surf sites include those at river mouths where changes in sediment outflow can alter morphology of the area; thus what happens inland can directly affect the sites. The natural watershed of San Mateo Creek, California, is a highly publicized example, where a naturally-occurring outflow of cobblestones geologically creates several world-class surf sites, collectively known as Trestles, and organizations such as the Surfrider Foundation are protesting the development of a toll road which will alter the outflow of the watershed (Surfrider Foundation, 2016b; Sustainable Surf, 2016; Nelsen *et al.*, 2007). Sites are also sensitive to changes, and features that can slow or obstruct ocean swells from traveling to a given coast. Offshore develop-ments, such as artificial reefs, wind farms and Wave Energy Converters (WECs) can block or slow waves from reaching sites (Butt, 2010). In consideration of these examples, surf site boundaries can be extended well beyond the immediate area to include the wider terrestrial and ocean natural systems, and this concept can be extended further to include the winds and weather systems that produce the waves. Consequently, surf site integrity is intrinsically tied to the implications of climate change and sea level rise.

Surfing habitat

Surf sites are part of a wide and encompassing system of natural processes. Sustainable Surf (2016) defines *surfing habitat* to include waves, oceans, marine animals (fish, seals, whales, sea birds), coral reefs, rocky reefs, ecosystem flora and fauna (plankton, kelp), and watersheds on land.

Direct human impacts on surfing habitat include threats identified to have a multiplier effect on the environment, such as over-fishing, urban pollution (sewage, urban runoff, industrial discharge), sedimentation, marine debris, coastal development, oil spills, watershed land-use change (Sustainable Surf, 2016). In the face of these issues, Buckley (2002) proposes that surf sites, depending on how commercial surf tourism is managed, are jointly vulnerable to major environmental impacts and hold the potential to help with the conservation of native habitats and traditional cultures.

Surf habitat conservation

Conservation is in effect the sensible and careful use of natural resources by humans whereby individuals are concerned with using natural areas in ways that sustain them for current and future generations of human beings and other forms of life (Miller, 2006). As the concept of coastal conservation often includes stakeholder use and community involvement with the ultimate aim of maintaining environmental integrity, significant to the implementation of conservation ideals is the proactive management and use of various coastal planning approaches (Kay and Alder, 2005), and these actions are most effective when accounting for the environmental capital of a given area. Thus, when placing sites in the context of protection or conservation, we must account for a number of sensitivities which may determine the design or structure of the management plan (Barrow, 2005). Biologist R. Ritchie explains:

> The conservation of surfing sites is much like conserving elephants; it requires the protection of habitat which encompasses not only a large area but also any number of other resources and species.... Therefore, conservationists who seek the protection of habitat like the idea of protecting surfing areas for this reason.
>
> (in Martin, 2013: 31)

The recognition of surfing areas as a coastal resource worthy of protection is a relatively recent development sparked in part by the prolific growth of domestic and international surf travel which has spread surf tourism to cities and rural areas around the world. Surf tourism has awakened coastal communities and local and regional governments to the significance and consequences associated with the loss or degradation of the resource. Only recently has research validated the habitat importance of surf sites when conducting Environmental Impact Assessments in coastal projects (Butt, 2010; Scarfe *et al.*, 2009). Butt (2010) identifies a number of ways in which waves can be adversely affected or lost, including the construction of solid structures (which are common and permanent), dredging river mouths and canals, chemical pollution and sewage, oil spills, nuclear waste, and litter and marine debris, in addition to problems with access. For example, water quality is widely understood as foundational in the health of surf habitats and the surfers who visit them (Butt, 2010, 2011; Martin 2013; Martin and Assenov, 2012, 2014; Ryan, 2007). In terms of conservation ecology, Ritchie (in Martin, 2013: 33) suggests:

> We must consider that surfers require clean water and beaches, and water quality is a serious issue – if you get sick surfing an area you will likely not come back – nobody wants to surf or vacation at a polluted area.

Research in conservation action planning includes a great number of considerations and approaches in order to address human impacts and other issues affecting the resource base, including rapid assessment strategies (TNC, 2007), and surf site research is no exception. In order to segment research data, indices can be employed to identify assessable qualities or attributes that contribute to conservation for any given surf location. By categorizing sets of indicators to form social, economic, environmental, and governance indices, data can be applied quickly to distinguish key issues at threatened sites. For example, the Surf Resource Sustainability Index (SRSI) (Martin, 2013; Martin and Assenov, 2013, 2014, 2015) employs a multidimensional approach by placing sustainability indicators into qualitative and quantitative modules for analysis, serving as a theoretical compass pointing at surf habitat conservation issues.

The demarcation of surf sites

A contemporary and conceptual recognition of surf sites first arose without the consideration of the physical boundary or demarcation of the surfing area per se; rather plaques and statues were displayed at sites to recognize local cultural icons or to encourage tourism, such as in Freshwater Beach (Australia), Pipeline and Waikiki (Hawai'i, USA), Santa Cruz (California, USA), and Uluwatu (Bali, Indonesia) (Farmer and Short, 2007). While all these sites are clearly acknowledged to have a strong association with surfing, none had formal mechanisms in place capable of protecting or enhancing the site for surfing. For this to occur, as well as visual recognition, a reserve system can be employed to identify and protect iconic surfing sites (Farmer and Short, 2007).

Some of the earliest demarcations of surf sites were in ancient Hawai'i, where sociopolitical management systems emphasized the significance, use, and physical boundaries of sites. As revealed in an interview with K. Koholokai (Martin, 2013: 34), stories and legends of the Hawaiian surf sites lend credit to the contemporary concept of the surfing reserve. The native Hawaiians have been surfing these sites for many centuries:

> Ancient surfing sites like *Ku'emanu Heiau* adjacent to Kahalu'u Beach Park [Kona, Hawai'i] and *Hale'a'ama Heiau* at Kamoa Point [Kona, Hawai'i] (today called the 'Lyman Point break') were afforded a type of protection according to traditional Hawaiian culture. Since ancient *He'e Nalu* (Hawaiian surfing) was a religious expression especially for the Ali'i or chiefly clans, it required surfing protocols of *Pule* (prayers), *Oli* (chants), *Ho'okupu* (offering), and *Kapu kai* (ceremonial sea bath). Surf sites like Ku'emanu and Hale'a'ama Heiau were several of the many physical and spiritual sites set aside for *He'e Nalu* (surfing). There were *ili* (strips of land) within an *Ahupua'a* (land division units) that was divided into smaller parcels of land like *Mala'ai* (plantation or gardens) and even *ili Kupono* or *ili Ku* (reserved chief lands) and *ili lele* (small parcels of land here and there). Kamoa Point is an *ili Ku* land division unit set aside for surfing and other sports activities, so *ili Ku* was not subject to tax or tribute by a *Konohiki* (landlord) of the *Ahupua'a*.

Although the contemporary lifestyle, sport and industry behind surfers and surfing have spread worldwide, Short and Farmer (2012) note that surf breaks are the very core of this activity and have 'largely been taken for granted.' They point to surf tourist destinations where the expanding surfing sector has done little to prevent the loss or contamination of sites: for example, the adjacent environment has not been protected from inappropriate development. Key issues include surf sites being overwhelmed by development, population pressures, and the associated shadowing, pollution, sewerage, stormwater (Short and Farmer, 2012), and beach erosion because of the cutoff of the supply of sand (AECOM, 2010).

Farmer and Short (2007) note that surf sites have physical and social dimensions which include the beach and adjacent surf zone. They note that surf sites include not only the physical features of the marine and coastal zone which intrinsically enhance aspects of the surfing experience; they may include structures such as surf clubs. Social attributes include the surf site history or places considered sacred by surfers for a particular reason (Martin and Assenov, 2014).

Surfing reserves

While the conservation of coastal areas has a long history in many regions around the world, the protection and management of surf sites is a relatively recent phenomenon. The surfing reserve concept opens a new dialogue for the theoretical, practical, and political applications of surf site recognition and conservation. The first ever surfing reserve was formed in 1973 at Bells Beach, Victoria, Australia and serves as a milestone in surf conservation history. The original legislation was land-based, essentially protecting only the foreshore and terrestrial park area (FFLA, 2010).

Coastal conservation favours human use and interaction as integral to the sustainability of a given area and many coastal zones are set aside as parks and reserves intended to serve as habitat for wildlife, provide space for recreation and tourism, access to fishing grounds, or other purposes aimed at the conservation of natural resources. Broadhurst (2001) points out that parks and reserves have different meanings in different circumstances, the former suggesting some return of benefit to the user, the latter being concerned more with conserving the potential to provide a return for future generations. However, Kay and Alder (2005) suggest that the ability of conservation areas to meet the multiple-use demands of coastal users while providing for conservation is questioned by environmental preservationists who seek multiple-use as only a trade-off between economic development and preservation.

Broadhurst (2001: 145) asks, 'If we designate a place as special, does that mean that other places are not special?' In theory, the conservation of special places exists only in the human mind, as an abstract concept aimed at changing people's behaviour or the side effects of their behaviour. In practice, for conservation to work, people must first agree to have a conservation area, and what rules to apply, and the stakeholders must understand what to do or what not to do in the context of a wider and variable chain of events (Anthoni, 2001). Thus, while one particular area may be resistant to various human or natural impacts that cause environmental change, another area may be highly susceptible, and the designations of environmental zones need to be site-specific and take into account a wide range of criteria (Broadhurst, 2001). Concerning vulnerable and biodiverse marine environments, Jessen *et al.* (2011) identify that sustaining ocean health requires ecosystem-based approaches to management and that Marine Protected Areas (MPAs) are a central tool in this context. In a broad sense, MPAs include areas of the coastal zone or ocean conferred a level of protection for the purpose of managing use of resources and ocean space, or protecting vulnerable or threatened habitats or species (Dimmock, 2007).

The most comprehensive strategy to date for the direct protection of surf sites is the concept of the 'surfing reserve' (Farmer and Short, 2007; Short and Farmer, 2012), including representational announcements by the local community. The promulgation or 'symbolic declaration' of surfing reserves is imperative in recognizing surfing activities as vital to a particular area, including the socioeconomic and cultural value of surfing, wherein the surfing community is interested in developing a long-term management plan in conjunction with the local land management authority (Lazarow, 2010).

A surfing reserve is designed to formally recognize surfing sites and in doing so provide a focus for the ongoing protection of those sites and to assist in the concerned management and development of the adjacent land area; it is a proactive step towards surf site conservation and represents a mechanism to redress the 'casual attitude' of surfers to their surf breaks (Short and Farmer, 2012). Justification for surf break protection through a reservation system can take into account an ever-increasing and mobile surfing population, unrelenting environmental and development pressures in the coastal zone, and the 'less than impressive record of mass tourist development and destruction that has followed on from surf break discoveries in many third world locations' (Lazarow, 2010: 265).

Short and Farmer (2012) note that surfing reserve boundaries vary considerably from one site to another, ranging in size from just a few hundred metres of coast to several kilometres. Sites should extend from the shoreline at least 500 to 1,000 metres out to sea to make sure the breaks themselves are included. They provide examples in Australia where the reserves include the surf breaks, the coast, and the surrounding ocean, and range in extent from 600 metres of coast and 50 hectares in size to over seven kilometres of coast and 400 hectares. While surfing reserves may not have any direct bearing on adjacent land use, they may provide a substantial support in the debate about adjacent land use and development (Short and Farmer, 2012).

Stakeholder dimensions

Economic Linkages

Understanding the broad scope and relationships among surf resource stakeholders is a relatively new endeavour. Researchers and economists have only recently begun to investigate the value of waves and identify the significance of various stakeholder groups. Most evident are the individual surfers who bring money to local businesses and the wider coastal economy when they go surfing (for example, making local purchases of provisions and petrol). While surfers are an obvious stakeholder group, their capacity goes beyond riding the waves and includes their employment in various businesses and surf-related industries intrinsically tied to a particular coastal area. Butt (2010) identifies that surf resource stakeholders include surfers and other members of the community who own or work in surf-related establishments where the visitors spend their money, including surf shops, surfboard manufacturers or surfing schools. Similarly, there are businesses that may derive income based on the existence of a good surfing wave in their town through extrinsic and less obvious sources, such as airlines, rental car companies, petrol stations, restaurants and bars, etc. For instance, the AEC Group (2009) found that surf businesses on the Gold Coast, Australia created local employment for a number of high-skill occupations tangentially connected to the resource, including graphic designers, filmmakers, journalists, web designers, legal and finance professionals, as well as the more obviously related areas of surfboard shaping, clothing and hardware design, surf schools, educators, and surf media. Although non-surfers, such as hotel employees, managers, shop owners, politicians, or anybody else with a relationship to the site, may not have a direct stake in riding the waves, they can have indirect stakes, including employment in service industries, and other social and economic interests (Butt, 2010).

Another dimension of stakeholders in surf sites are interests connected with surfing events. O'Brien (2007) notes that impacts on host communities and linkages among stakeholders include contest sponsors, surf shops, hotels, advertisers, banks, stores, restaurants, and bars, resulting in short and long-term benefits and enhanced business relationships. He notes that key sectors include surfing hardware, surf accessories and services, hospitality accommodation, and event-related infrastructure. Additionally, in order to set up and run the event, local suppliers provide infrastructure, such as scaffolding, tents, public address systems, trophies, prizes, and T-shirts, as well as services, such as 'qualified judging, travel, accommodation and hospitality solutions, media and photographic services, and entertainment venues for event augmentations' (O'Brien, 2007: 152).

Stakeholders and surf system sustainability

Martin and Assenov's (2012) review of surf tourism research suggests a need to define the complete system boundaries of surf sites, including the significance and activities of new regional and demographic markets, surfwear manufacturers and the sponsorship of surf events, cultural shifts in the surfing subcultures, and the impacts of technology and coastal engineering innovations such as artificial surfing reefs. While these topics are of growing interest in the academic community, published research attesting to the physical and human 'surf system' as a holistic spectrum of social, economic, and environmental criteria and implications for sustainability is limited. To address these concepts, sustainable surf site policy and management must attend to various local ecosystems as a range of complex, diverse yet integrated components with essential linkages spanning people, places, and impacts on a vulnerable resource base consisting not only of the water, waves, reefs, and coastal morphology, but also of the coastal users as stakeholders, local infrastructure, and economy (Martin, 2013).

The argument that waves are resources, and that a wide range of stakeholders are players in their sustainability, has only recently appeared in academia, particularly as a result of graduate research and the not-for-profit sector (Martin and Assenov, 2012). For example, Butt (2010) (in a report commissioned by Surfers Against Sewage) suggests that the world's coasts and waves are indeed natural resources and can be used to benefit everyone in a sustainable and stable way. He notes that the wider consequences of degrading or destroying surf breaks are not well understood and may seem inconsequential, but the implications should be taken seriously: 'We don't know where the threshold is; we don't know how much we can modify the system before it goes out of balance' (Butt, 2010: 45).

In the wide view, natural resources are a component in 'natural capital', which includes the services with which nature provides us and other species, such as those that sustain life and support our livelihoods and economies (Miller, 2006). It is in this context that Surfing Capital (Lazarow et al., 2007, 2008) brings the argument of natural capital into the context of surfing through itemizing the natural and human impacts relative to wave quality and frequency along with environmental and experiential dynamics. Under the framework of Surfing

Capital, Lazarow *et al.* (2007, 2008) draw a list of direct stakeholders that includes biologists, climate change specialists, coastal developers, engineers and managers, environmentalists, legislators and politicians, social scientists, a wide range of amenity stakeholders in the built and natural environment, and various stakeholders in issues of public access and safety covering both public and private property. Thus, the sustainability of the integral surf system relies on the ability of diverse stakeholders to engage in dialogue and education covering such issues as the elucidation of surf sites as emergent and dynamic coastal resources. These sites are increasingly being recognized as natural capital, the sustainability of which can only be achieved by their wise and careful management. Miller (2006: 8) places the concept of managing natural capital in the context of one's own economic integrity: 'Protect your capital and live off the income it provides. Deplete, waste, or squander your capital, and you will move from a sustainable to an unsustainable lifestyle.'

Surfers as resource stakeholders

Surfing is an important recreational and cultural use of the coastal zone and surfers are an important coastal stakeholder group; they have strong cultural passion and sense of ownership of their surf spots as 'natural cultural resources' ASBPA (2011). Counter to the stereotype of surfers as unemployed beach bums, experienced surfers often have college degrees and are mostly in the upper middle-class income bracket (Nelsen *et al.*, 2007). However, surfers constitute a coastal interest group that has historically been ignored in coastal management (Scarfe *et al.*, 2009). Butt (2010) writes extensively on the role of the surfers as a significant stakeholder group directly affected by the integrity of surf site sustainability. He notes that if a surf site is destroyed, polluted, or degraded for some reason, the surfers in the town will not only suffer because they will be unable to surf it, but they might also suffer because their jobs depend on that wave bringing money-spending tourists into town. For example, a sudden loss of revenue occurred to the community at Mundaka, Spain, where a coastal dredging project degraded a world-class wave (Murphy & Bernal, 2008).

The role of surfers is essential when considering the identification, preservation, or mitigation of surfing resources in coastal planning and project development (ASBPA, 2011). Accordingly, by engaging surfers, ideas or concerns can be addressed early in the coastal management process. Scarfe *et al.* (2009) suggest that as the social, economic, and environmental benefits of surfing breaks are realized, surfers are increasingly integral players in coastal resource management. For example, surfers can pinpoint areas of special interest that developers should avoid and they have a role to play in promoting the following basic principles: conserving and enhancing natural and cultural heritage, sustainable use of natural resources, understanding and enjoyment of the environment through recreation, and the sustainable social and economic development of local communities (Butt, 2010). Surfers are also core stakeholders in the case of urban sites which they identify as their local breaks and at sites where good wave quality attracts locals and travelling surfers alike, including world-renowned iconic breaks (Short and Farmer, 2012). In terms of education and public awareness, surfers suggest that knowledge empowers the public through promoting relevant issues (Martin, 2013; Martin and Assenov, 2014), and studies by Lazarow (2010) suggest that in the long run, educating the surfing community and public on the importance of sites and their protection is crucial to surf site protection. The need for public awareness is being met by the rise of grassroots surf organizations, which have increased significantly in recent years.

Grassroots surf organizations

ASBPA (2011) highlights the fact that surfers are becoming increasingly organized as stakeholder groups in protecting existing surf spots and supporting coastal management policies that take into consideration social, economic, and environmental implications. At the local, regional, and national not-for-profit level, some well-known examples include Save the Waves Coalition, SurfAid, Surfers Against Sewage, Surfrider Foundation, Surfers Environmental Alliance, Waves for Development, and Wildcoast.

Surfers may also form local and regional boardriders and lifesaving clubs. These organizations are usually based at or centred on surf sites, and form independent stakeholder groups. Augustin (1998) notes that these clubs can unite to form national federations, and play a vital role in the promotion of surfing. They may also help to inspire synergies among surfing sponsors, the media, surfers, and local communities. Surf lifesaving clubs may form independently or under the auspices of local or regional governments, and can become grassroots stakeholder groups directly related to site integrity in terms of community, education, and safety (AECOM, 2010). The public standing of lifesaving clubs is very high and they usually have good access to local and state government (Martin and Assenov, 2014).

Surf tourism stakeholders

In terms of surf tourism, Buckley (2002) offers four interconnected groups of stakeholders which influence the role of surf tourism in sustainable development. They include individual surfers, the commercial and competitive tour operators, local residents, and government officials. He notes that the ethics among surfers form a complex fabric of stakeholder responsibility along with the desires and codes among tour operators, the traditional and modern perspectives of host communities, and the requirements of governments.

Bearing in mind the global surf tourism industry, surf resource sustainability is of growing significance to a wide range of stakeholders in very different socio-economic and cultural settings (Martin and Assenov, 2012). The most obvious differentiation is between urban 'surf city' economies in the developed world, for example the bustling Gold Coast, Australia, or San Sebastian, Spain, and rural island settings in developing countries, such as the Mentawai Archipelago,

Indonesia, and Lobitos, Peru. In the case of the former, Surf Cities are coastal communities where surfing plays an instrumental role in the character and fabric of the community and tourism industry. The World Surf Cities Network (2016) defines a Surf City as an urban area where surfing, surf culture and employment in surf industries are relevant to the economic, social and cultural base of the city and the surf industry is formally recognized by the city government. Tangible elements include not only physical location, population, natural resources, and the manufacturing industry, but also services and culture. Services comprise those relevant to surf tourism, surf retail, surf schools, surfing events and competitions, surf training, surf media, and the real estate market. Culture takes account of the surfer population, surfing associations, surf culture events and history, and recognition by the city (World Surf Cities Network, 2016).

In the case of rural coastal communities in the proximity of surfable waves, these communities inevitably became key stakeholders in surfing resources, with various positive and negative outcomes (Ponting and O'Brien, 2014, 2015; Towner, 2016). Apart from the negative effects and influences brought by the unplanned and rapid advance of the surf tourism industry in various locations around the world, positive outcomes include surfer-volunteerism programmes in community outreach, environmental health, and entrepreneurship empowerment (Waves for Development, 2016). Similarly, SurfAid International (2016) is a well-publicized example of a not-for-profit organization focused on community development through improving the health, well-being and self-reliance of people living in isolated regions, particularly in Indonesia. Thus, the concept of the surf tourism stakeholder broadens to include those who provide, receive, and benefit from community-based health and education in these regions (Ponting and O'Brien, 2015).

Traditional resource custodians

Traditional resource custodians at surf sites include host communities, such as fishing villages on islands and in developing countries which may have long-standing access rights and interaction with coastal resources. Previous to the global exploration and exploitation of surfing resources in such areas, the significance and value of surf resources were typically not recognized by local communities. As a result, with the arrival of the global surf tourism industry, including groups of travelling surfers on land and by boat, rural host communities had no experience in managing these resources and were unprepared for the social and economic implications and impacts. Buckley (2002) relates that commercial surf charter boats and land-based surf camps have typically operated as enclaves with little meaningful interaction with local host communities. Ponting (in Martin, 2013: 44) identifies the contrast between surf tourism operators and traditional resource custodians: 'The million-dollar boat and the impoverished community'.

Research by Ponting *et al.* (2005) indicates that unregulated free-market approaches to surf tourism development in less developed regions alienate local people as a single and comparatively powerless or displaced stakeholder group amongst many others. Consequently, local people are often the last to benefit from economic development based upon the exploitation of their resources, yet shoulder the bulk of negative impacts and feel resentment. Indigenous communities risk exclusion from the surf tourism economy (Ponting *et al.*, 2005), and surf tourists may miss out on the opportunity of important cultural exchange to add value to their experience (O'Brien and Ponting, 2013; Ponting and O'Brien, 2014).

A. Abel (in Martin, 2013: 44) explains that in the case of Papua New Guinea (PNG), host communities can be seen as 'traditional resource custodians', a more holistic concept than the contemporary concept of 'land owners'. The failure of other stakeholders to recognize this distinction left locals marginalized in the use of their coastal resources by surf tourists. As President of the Papua New Guinea Surfing Association, Abel has worked to educate and empower local communities through a consultation process aimed at social and economic sustainable development. Abel's approach helps indigenous communities to embrace the benefits of surfing waves as a renewable resource on their own terms, employing practical methods such as limiting the number of users of sites in order to manage social and environmental impacts, while providing economic benefit to the community and a unique cultural experience for the surf tourists. Abel (in Martin, 2013: 44) explains:

> We are building a new conceptual 'bottom-up' model to surf tourism, where indigenous communities manage their resources in a sustainable fashion as stakeholders – and this has even helped to promote protection of the surf reefs through abandonment of harmful fishing practices which once used dynamite and cyanide.

O'Brien and Ponting (2013) note that surf management plans have been developed and put in place to solve a variety of issues in PNG where reefs are owned by local villages or clans and the rights to natural resources do not end at the high-water mark as they do in most countries; rather their traditional grounds include the reefs where the surfing activities now take place. Thus, in the case of a commercial surf tourism operation, which utilizes an area to conduct business, it is appropriate for the traditional resource custodians of the reefs to benefit. However, while extracting 'reef fees' for fishing has some cultural heritage in many indigenous communities, managing reefs for surfing is somewhat of a foreign concept to such communities as revealed in the following interview conducted by O'Brien and Ponting (2013) in PNG:

> This was a resource that they didn't realize they had. They had the potential to develop, manage, promote, and at the same time, derive a sustainable source of income without denigrating their day-to-day way of life, their culture, or their heritage.
>
> (2013: 168)

At the time of writing, PNG's surf tourism sector serves as the only example in the world of a formalized attempt by indigenous surf resource custodians to collaborate with stakeholders to sustainably manage surf tourism resources and activities through a community-centred strategy. This approach engages resource owners in planning acceptable use of their surfing resources and appropriate compensation (O'Brien and Ponting, 2013).

Fiji serves as another case study in the Asia-Pacific. Ponting and O'Brien's (2014) research notes that traditional fishing grounds have been a source of controversy dating back to the colonial era, and this has been exacerbated by the development of the lucrative commercial surf tourism industry, which consists of as many as 75 tour operators at 120 surf sites. Recent changes in access to these resources by the government have caused tensions to escalate among individuals and communities and created an environment of social and political uncertainty. At the time of writing, new open-access policies to Fijian surf sites have come at the cost of 'de-territorialization' of customary resources and mark a transition from communally-owned common pool resources – and the impacts to sustainability are yet to be determined (Ponting and O'Brien, 2014).

To address these issues, management strategies allied to differing culturally bounded property rights need to be developed accordingly; and Ponting and O'Brien (2014, 2015) suggest that regulatory philosophies and frameworks should consider compensating indigenous resource custodians for the use of their reefs and fishing areas. Their research in PNG (O'Brien and Ponting, 2013), Fiji (Ponting and O'Brien, 2014), and the Mentawai Islands (Ponting and O'Brien, 2015), as well as Towner's (2016) work in the Mentawais, highlights the integral juxtaposition of sustainability and surf tourism; it may also exemplify how the development of surfing activities at the village level can foster better management of surf sites by indigenous communities and other stakeholders through insightful planning for sustainability and increased opportunities for local communities to share in the benefits derived from surf tourism.

Interdependence of stakeholders

Two paradigms coexist when looking at the contemporary understanding of surfing sites in the social sciences – the global value perspective of the surfing industry alongside the value attributed to specific surfing locations by individuals and local communities. Given the enormous reach of the global corporate surfwear and equipment industries, combined with the increase in the number of individual surfers and surfing communities in the world who contribute to the visitation of sites, these factors encompass many facets of tourism, direct and indirect values, and stakeholder linkages and engagement. While relevant market values are reasonably easy to measure through, for example, domestic and international tourism receipts from surfing schools, camps, and events, the nonmarket values such as the economic benefits of regional and national image, sociocultural aspects, physical fitness, and psychological well-being are more difficult to measure. Nevertheless, nonmarket values touch the lives of millions of surf resource stakeholders in coastal areas across the world.

There has been relatively little research which investigates surfing sites in a holistic systems context, whether in terms of the individual, society, the economy, or focused on the conservation of the natural environment. The study of surf resource systems boundaries theoretically highlights the evidence-based role of the environmental and social sciences in the management of coastal surfing resources, setting the stage for the use of new and interdisciplinary methods in surfing and sustainability research.

References

AEC Group Ltd. (2009). GCCC surf industry review and economic contributions assessment: Gold Coast City Council. Gold Coast: Gold Coast City Council.

AECOM Australia Pty Ltd. (2010). Beach sand nourishment scoping study: Maintaining Sydney's beach amenity against climate change sea level rise. Sydney: Sydney Coastal Councils Group (SCCG).

Anthoni, J. F. (2001). Conservation principles 3: The spiritual dimension. Seafriends Marine Conservation and Education Centre. Leigh, New Zealand: Retrieved from www.seafriends.org.nz/issues/cons/conserv3.htm#spiritual dimension (accessed 1 October 2016).

ASBPA. (2011). White paper: Surfers as coastal protection stakeholders. Fort Myers, FL: American Shore and Beach Preservation Association.

Augustin, J. P. (1998). Emergence of surfing resorts on the Aquitaine littoral. Geographical Review, 88(4), 587–595.

Barrow, C. J. (2005). Environmental management and development. Oxon, UK: Routledge.

Bicudo, P. and Horta, A. (2009) Integrating surfing in the socio-economic and morphology and coastal dynamic impacts of the environmental evaluation of coastal projects. Journal of Coastal Research, 56(2), 1115–1119.

Broadhurst, R. (2001). Managing environments for leisure and recreation. New York: Routledge.

Buckley, R. C. (2002). Surf tourism and sustainable development in Indo-Pacific islands: I. The industry and the islands. Journal of Sustainable Tourism, 10(5), 405–424.

Butt, T. (2010). The WAR report: Waves are resources. Cornwall, UK: Surfers Against Sewage.

Butt, T. (2011). Sustainable guide to surfing (report). Cornwall, UK: Surfers Against Sewage.

Dimmock, K. (2007). Scuba diving, snorkeling, and free diving. In Jennings, G. (ed.), Water-based tourism, sport, leisure, and recreation experiences. Oxford: Elsevier. 128–146.

Farmer, B., and Short, A. D. (2007). Australian national surfing reserves rationale and process for recognising iconic surfing locations. Journal of Coastal Research, 50(SI), 99–103.

FFLA (Fitzgerald Frisby Landscape Architecture). (2010). Bells Beach Surfing Reserve coastal management plan, draft – 2010. Docklands, Australia: Surf Coast Shire.

GOP (Geological Oceanography Programme). (2013). Beach Systems. Retrieved from http://geology.uprm.edu/Morelock/beachsys.htm (accessed 1 October 2016).

Jessen, S., Chan, K., Côté, I., Dearden, P., De Santo, E., Fortin, M. J., Guichard, F., Haider, W., Jamieson, G., Kramer, D. L., McCrea-Strub, A., Mulrennan, M., Montevecchi, W. A., Roff, J., Salomon, A., Gardner, J., Honka, L., Menafra, R., and Woodley, A. (2011). Science-based guidelines for MPAs and MPA networks in Canada. Vancouver: Canadian Parks and Wilderness Society (CPAWS). Retrieved from http://cpawsbc.org/upload/mpaguidelines_bro_web_3.pdf (accessed 1 October 2016).

Kay, R., and Alder, J. (2005). Coastal planning and management. New York: Taylor & Francis.

Lazarow, N. (2010). Managing and valuing coastal resources: An examination of the importance of local knowledge and surf breaks to coastal communities. PhD dissertation, Australian National University, Canberra.

Lazarow, N., Miller, M. L., and Blackwell, B. (2007). Dropping in: A case study approach to understanding the socioeconomic impact of recreational surfing and its value to the coastal economy. Shore and Beach, 75(4), 21–31.

Lazarow, N., Miller, M. L., and Blackwell, B. (2008). The value of recreational surfing to society. Tourism in Marine Environments, 5(2–3), 145–158.

Martin, S. A. (2013). A surf resource sustainability index for surf site conservation and tourism management. PhD dissertation, Prince of Songkla University, Hat Yai, Thailand.

Martin, S. A., and Assenov, I. (2012). The genesis of a new body of sport tourism literature: A systematic review of surf tourism research (1997–2011). Journal of Sport and Tourism, 17(4), 257–287. doi: 10.1080/14775085.2013.766528.

Martin, S. A., and Assenov, I. (2013). Developing a surf resource sustainability index as a global model for surf beach conservation and tourism research. Asia Pacific Journal of Tourism Research, 19(7), 760–792. doi: 10.1080/10941665.2013.806942.

Martin, S. A., and Assenov, I. (2014). Investigating the importance of surf resource sustainability indicators: Stakeholder perspectives for surf tourism planning and development. Tourism Planning and Development, 11(2), 127–148. doi: 10.1080/21568316.2013.864990.

Martin, S. A., and Assenov, I. (2015). Measuring the conservation aptitude of surf beaches in Phuket, Thailand: An application of the surf resource sustainability index. International Journal of Tourism Research, 17(2), 105–117. doi: 10.1002/jtr.1961.

Miller, G. T. (2006). Environmental science: Working with the Earth. Belmont, CA: Thompson Learning.

Murphy, M., and Bernal, M. (2008). The impact of surfing on the local economy of Mundaka Spain. Davenport, CA: Save the Waves Coalition.

Nelsen, C., Pendleton, L., and Vaughn, R. (2007). A socioeconomic study of surfers at Trestles Beach. Shore and Beach, 75(4), 32–37.

O'Brien, D. (2007). Points of leverage: Maximizing host community benefit from a regional surfing festival. European Sport Management Quarterly, 7(2), 141–165.

O'Brien, D., and Ponting, J. (2013). Sustainable surf tourism: A community centered approach in Papua New Guinea. Journal of Sport Management, 27(2), 158–172.

Ponting, J., and O'Brien, D. (2014). Liberalizing Nirvana: An analysis of the consequences of common pool resource deregulation for the sustainability of Fiji's surf tourism industry. Journal of Sustainable Tourism, 22, 384–402.

Ponting, J., and O'Brien, D. (2015). Regulating 'Nirvana': Sustainable surf tourism in a climate of increasing regulation. Sport Management Review, 18, 99–110.

Ponting, J., McDonald, M. G., and Wearing, S. L. (2005). De-constructing wonderland: Surfing tourism in the Mentawai islands, Indonesia. Society and Leisure, 28(1), 141–162.

Ryan, C. (2007). Surfing and windsurfing. In G. Jennings (ed.), Waterbased tourism, sport, leisure, and recreation experiences. Oxford: Elsevier. 95–111.

Save the Waves Coalition. (2010). Malibu world surfing reserve. Davenport, CA: Author.

Scarfe, B. E., Healy, T. R., Rennie, H. G. and Mead, S. T. (2009). Sustainable management of surfing breaks – an overview. Reef Journal, 1(1), 44–73.

Short, A. D., and Farmer, B. (2012). Surfing reserves: Recognition for the world's surfing breaks. Reef Journal, 2, 1–14.

SurfAid International. (2016). About SurfAid. Retrieved from www.surfaid.org/about (accessed 1 October 2016).

Surfrider Foundation. (2016a). Environmental policies. Retrieved from www.surfrider. org/pages/environmental-policies (accessed 1 October 2016).

Surfrider Foundation. (2016b). Save Trestles: Stop the 241 toll road extension. Retrieved from http://savetrestles.surfrider.org (accessed 1 October 2016).

Sustainable Surf. (2016). Loss of surfing habitat. Retrieved from http://sustainablesurf. org/eco-education/loss-of-surfing-habitat/ (accessed 1 October 2016).

TNC (The Nature Conservancy). (2007). Conservation action planning: Developing strategies, taking action, and measuring success at any scale. Arlington, VA: Author. Retrieved from www.conservationgateway.org/Documents/Cap%20Handbook_ June2007.pdf (accessed 1 October 2016).

Towner, N. (2016). Community participation and emerging surfing tourism destinations: A case study of the Mentawai Islands. Journal of Sport & Tourism, 20, 1–19.

United Nations. (1987). Report of the world commission on environment and development: Our common future. Annex doc. A/42/427. New York: United Nations.

Waves for Development. (2016). About waves. Retrieved from www.wavesfordevelopment. org/about/#.V-u1UCQnVxs (accessed 1 October 2016).

World Surf Cities Network. (2016). About world surf cities network. Retrieved from www.worldsurfcitiesnetwork.com/en/wscn-en/about-us (accessed 1 October 2016).

Part III

Technology, industry, and sustainability

3 Surfing in the technological era

Leon Mach

Introduction: the technological era

Surfers, surf historians, and sociologists tend to look back on the history of surfing and categorize different time periods into eras or epochs. This narrative begins with the foundational era (fifth to eighteenth century), when surfing itself was born from the cradle in Polynesia and nursed into one of the main pillars of the ancient Hawaiian societal structure (Finney and Houston, 1966; Laderman, 2014; Moore, 2010; Scures, 1986; Walker; 2011; Westwick and Neushul, 2013). Following this was the era of decline, characterized by European contact and settlement in Hawai'i beginning at the end of the eighteenth century, which led to the decimation of the local population and the suppression of many religious and cultural activities (Marcus, 2009; Walker, 2011; Westwick and Neushul, 2013). The Western appropriation era followed, when surfing was picked up and adopted by pockets of people in coastal enclaves in the West before captivating a mainstream audience following the world wars (Irwin, 1973; Westwick and Neushul, 2013). Then came the countercultural epoch, when 'soul' surfers attempted to differentiate themselves from the mainstream by travelling the world to escape the growing crowds and dosing themselves with drugs in an attempt to reinsert a freedom and artistic expression into the activity (Lawler, 2011; Ford and Brown, 2006). It is then often said that surf culture entered the commodification/professionalization era, when professional surfing became a job and was judged on standardized criteria and surf apparel industry and tourism providers became big businesses (Ford and Brown, 2006; Laderman, 2014; Westwick and Neushul, 2013). Certainly, era categorizations can mask cultural tensions and contradictions in each period discussed, but on net, they help a culture to make sense of its evolution and underlie the cultural narrative that situates the present in a coherent historical context.

It is always most difficult to diagnose and categorize the present. Though the topic has eschewed much discussion or attention, this chapter makes the case that we have entered the technological era of surfing – meaning the most significant differentiating characteristic of the modern surf period permeating the culture is the common acceptance of new technological inputs and the feverous pace at which the surf industry accelerates and extends the technological arsenal available to the contemporary surfer. Surfers today have access to fully fore-casted surf conditions on the Internet, as well as real-time cameras pointed at hundreds of surf breaks (and counting) around the world; a gaggle of board designs and materials (most of which are extremely toxic) to choose from; per-sonal waterproof and aerial image capturing devices; planes and helicopters to get to the surf (energy and carbon intensive); jet skis (also energy and carbon intensive) to propel them into giant waves; wetsuits that make arctic surfing pos-sible; and even manmade waves to surf both in and out of the ocean. In short, surfing has never before been more dependent on technological inputs and the surf industry's economic bottom-line never more closely correlated to extending the technological frontier and achieving higher levels of efficiency (whether that be in terms of travel, equipment design and production, or wave forecasting). The infiltration of high-technology into the surf world has undoubtedly been an ongoing social experiment, but no one fully comprehends where techno-surf culture is heading. Or perhaps more importantly, whether or not surfers even like the new world order being created with greater commitments to technological efficiency in what was once such a spiritual, stripped down, and back to the earth activity.

Using surfing as a lens, this chapter examines the promise and problems asso-ciated with technization of surfing to reinvigorate an old debate that seems to have been obliterated by a pervasive societal acceptance of technological pro-gress in many other spheres of existence. As far back as 1846, J. S. Mill said, '[h]itherto it is questionable if all the mechanical inventions yet made have light-ened the day's toil of any human being' and instead he suggested that 'they have enabled a greater population to live the same life of drudgery and imprisonment and an increased number of manufactures and others to make fortunes' (quoted in Noble, 1983: 17). More than a century later, it is critical to interrogate the role technological advance plays in guiding surf culture and as Mill suggested, this requires pondering whether or not technology liberates or enslaves the surfing population in service of its own perpetuation. The three integral questions asso-ciated with interrogating the technological era of surfing and examining the seeds of a transition to sustainability era are:

1   What technological dimensions categorize the technological era of surfing?
2   Who or what is guiding technological progress and towards what ends?
3   Is there evidence that surf culture is shifting towards a sustainability era?

The central argument is that without understanding the role of technology in environmental and social evolution, societies and cultures become vulnerable to technological determinism, whereby technology becomes an 'autonomous thing, beyond politics and society, with a destiny of its own which must become our destiny too' (Noble, 1983: 10). The danger in surf culture, as well as in other spheres of human existence, is when technological progress becomes both means and ends, effectively locking the activity, and the culture that has formed around it, in a paradigm of relentless acceptance of new technologies and greater efficiency for its own sake. This is a destructive path which requires great amounts of material throughput and toxic waste. This is not, however, the only path forward. When cultures recognize the challenges brought on by greater commitments to technological risk they can consciously develop social movements to challenging existing structures and usher in a new paradigm based on collectively derived normative ideals, such as with the animated anti-nuclear energy movement in the early 1970s US.

There is evidence that there is a growing undercurrent in surf culture attempting to forge a new and more sustainable surf future based on psychological and social well-being (Csikszentmihalyi, 1990; Flynn, 1987; Nichols, 2014), environmental sustainability (in sustainable product design, travel, and in the protection of coastal environments), and/or belonging to a community based on the shared tacit knowledge of the ecstatic experience of surfing (Stranger, 2011). This chapter seeks to explore the technological era and the evidence of the possible shift towards a sustainability era, which must be rooted in normative societal goals driven by human intention – this shift is not inevitable, but rather this outcome (or more likely this path) will take conscious effort and negotiation, multiple iterations, and constant adaptation.

In what follows, this chapter will first discuss a framework for defining and discussing technology in the present tense, which involves explaining technological determinism and its associated dangers before moving on to introduce a coevolving technology, environment, and society (TES) framework to situate an examination of the role technology could play in ushering in a new sustainability era. This chapter will then discuss the four dimensions of technology critical to understanding surfing and its evolution before concluding with examples that elements of the surf culture are beginning to challenge the core tenets of the technological era through attempts to meaningfully insert human agency in the process of guiding technological advance. In a sense, sowing the seeds for the transition towards a sustainability era in surfing, which could become a beacon of inspiration for other industries and lifestyle subcultures.

Ecologically connected imagery

One goal associated with digging into the technological components of surfing is to counter the prevailing imagery associated with representing surfing both directly to surfers and also to the outside world. This is the imagery used in the surf media (presented to surfers in films, magazines, and surf websites), in marketing campaigns staged by surf apparel and equipment manufactures, and also in popular media and marketing for an expansive array of goods, which depicts surfers as overtly ecologically connected (Lawler, 2011; Ormrod, 2007; Ponting, 2009). Surf apparel companies use slogans like 'live simply', 'live the search', and 'life is better in board-shorts' to propel an idea that surfing is raw and primitive – it is about man/woman communing with mother ocean and going to great lengths to satisfy this ancient desire. Images show surfers alone riding warm blue waves cresting over their heads, or camping simply on palm fringed shorelines eating fresh-caught fish and coconuts. Lawler (2011) argues that throughout the twenty-first century in America, surfing has symbolized every-thing counter to the production principle that defines industrial modernity because surf images conjure up a subconscious desire for a primitive past, a past when humans were more connected to the natural rhythms of the universe proper and were unencumbered by the trappings of the consumer society. The ability of surf imagery to achieve this sentiment explains why 'surfing sells' and is used as fodder for marketing campaigns to sell everything from Viagra to surf vacations to retirement planning services (Lawler, 2011).

Ponting (2008) deconstructed this surf image and argues that it is comprised of four general themes; tropical pristine environments, compliant, friendly locals, warm, uncrowded waves, and the lure of adventure, and these themes combine to show surfing as a passionate pursuit centred on living a stripped down life of dedication to moving with the ocean. In an era of anxiety, fragmentation, ennui, and moral relativism, surfing represents an Occam's razor approach to happi-ness, a pursuit that is simple and satisfying, irrational and hedonistic, gives meaning to life and binds the surf community in the shared knowledge of the ecstatic experience of wave riding (Stranger, 2011).

Behind the guise propelled by the media, however, there is a technological wedge being driven between man/woman and the waves, which has eschewed serious discussion and analysis. Human interactions with the natural world through surfing are becoming mediated more and more through a web of tech-nique, which Vanderburg (2005: 437) calls a technological 'cocoon around human life through which everything else is experienced'. The lone surfer getting head-high barrels in the images propagated by the surf industry (and other industries) was not camping there on the beach for months waiting to be blessed with a swell, he or she was hyper-responsive to surf reports, flew and then drove or jet boated directly to the spot, had the session filmed with a bunch of cameras, and then got out of there. It is now time to discuss a framework for critiquing technological advance and outline the technological dimensions on which modern surfing is reliant to begin to comprehend the implications of technological advance on surfing in the present tense so that surfers may use technology to guide the future towards a new epoch, rather than be guided by it in service of technological era mandates.

Defining technology in the present tense

The etymology of the word technology is born from the Greek word 'techne', which Aristotle discussed as the application of practical instrumental rationality to bring something into being – something equally capable of not being (Flyvb-jerg, 2001). In modern parlance, the term technology is most often used to describe an artefact (such as a surfboard or a computer) rather than a process, and technological progress is used as the chosen terminology for discussing the application of certain engineering processes embedded within a capitalist pro-duction system that brings new artefacts and modifications to existing objects into being (Rosenberg, 1972). There are thus, three critical components to interrogating technology – the processes that brings new artefacts, or efficiency improvements to existing artefacts into being (technique or technological progress), the artefacts themselves, and that which the objects enable users to do and achieve, as well as the impacts of these actions.

This section will discuss the first critical pillar, which is referred to by many philosophers of technology as 'technique' (Ellul, 1964; Noble, 1983; Vanderburg, 2005: 431). Technique is a complicated term as it describes the manner in which society becomes geared towards and put in the service of increasing commitments to technological advance. Technique also dictates the terms of this advance, which is calculable efficiency on route to the one best way to carry out any activity (Ellul, 1964; Vanderburg, 2005). It is similar to Schumpeterian 'creative destruction', whereby industries strive to increase efficiency incrementally (by trumping that which came before). Technique is the process that brings new artefacts into being, products that obviate the old, but it also guides the way society is organized to bring forth and consume these advances – efficiency and precision in production in order to produce technologies that enable tasks to be completed with calculably greater efficiency and precision. Historically, human cultures were once organized around religion, spirituality, animism, or some form of social contract, now however, with state sponsored acquiescence to the limitless potential of science in the furtherance of technique, society requires neither experience, the sacred, nor culture – it is a process of rational disenchantment – scientific epistemology becomes truth and that which cannot be quantified gets discounted (Vanderburg, 2005; Weber, 1976). In this vein, many suggest that technique has become deterministic, or an end in and of itself that is becoming further and further removed from human agency, oversight, and guidance (Ellul, 1964; Nye, 1998; Noble, 1983; Vanderburg, 2005). Technological advance is happening; humans simply must deal with it.

Determinism is by no means a new idea. Throughout the history of civilization there has always been staunch debate concerning whether stations, opportunities, and events in life should be considered pre-determined by a higher authority or whether humans have agency (free-will) in fashioning the direction of their own lives. Beyond religion, Marx famously postulated that 'the mode of production of material life determines the general character of the social, political and spiritual process of life' (quoted in Nye, 1998: 2). In religious determinism God dictates one's fate and for Marxists, the elite capitalist class determines the social relations that perpetuate the system.

Technological determinism is a theory that technology has become autonomous, that there is an ensemble of people, industries, and processes working to improve (based on measurable efficiency) the stock of our technological word and that these processes have become automated, rather than driven in the service of consensus-based human desires (Ellul, 1964). For example, most folks in high-income countries use small computers in their pockets they call smart phones, which can transmit information in nano-seconds through an array of applications. Before humans even really take time to evaluate what this access means, the dizzying processes of shrinking semi-conductors to allow for faster computation and to increase the stock of available apps continues for no other purpose than to earn profits through providing users with more information more quickly. Vanderburg (2005) describes technological determinism as a circumstance in which, 'technological growth feeds on the problems it creates for human life, society and the biosphere, thereby trapping us in a labyrinth of technique'. In relation to surfing, this theory would frame the discussion by suggesting that the problems associated with the growth of the surf population (related to tourism impacts, crowding generally, and the consequences of material production such as labour issues and environmental pollution etc.), are due to technological progress and that surf culture is tacitly putting its faith into technical end-of-pipe solutions (for example to create new waves, such as artificial reefs and surf-pools to deal with crowding and wave scarcity) to these challenges, which fail to get at the root causes. Technical solutions tend to mitigate, deflect, or further complicate, rather than resolve (Norgaard, 1994; Vanderburg, 2005).

Noble (1983) wrote a seminal piece titled 'Technology's Politics: Present Tense Technology', wherein he argues that technological determinism is a result of the failure of societies to view technology in the present tense. Humans tend to be distracted by the promises of technology and the potential future directions they 'might' lead us in, rather than question how that which is around us right now impacts us socially and psychologically, and the environmental outcomes our technological use facilitates. Technological determinism, Noble (1983: 10) says, is 'the domination of the present by the past – and technological progress – the domination of the present by the future' and he further suggests that these phenomena 'have combined in our minds to annihilate the technological present'. He suggests ever farther that 'this intellectual blindspot, the inability even to comprehend technology in the present tense, much less act upon it, has inhibited the opposition and lent legitimacy to its inaction' (Noble, 1983). Vanderburg, 2005: 449) sums up the failure of societies to evaluate technology in the present tense in the following passage:

> Every day people are promised new exciting gadgets, medical miracles, and scientific marvels, only to see them discarded shortly after. They are mesmerized by their new palmtop, the new bells and whistles on their cellphone, or a new 'smart' appliance, only to quickly discover that in the final analysis they make no real difference. These people need not make any value judgements. As long as they outgrow the previous fad along with everybody else they will be normal, and that is all they need to worry about.

To apply these ideas to surf culture suggests that surfers are simultaneously locked in a paradigm created by past technological advance and waiting for the future artefacts (and processes they enable) to be developed, and thus are rendered essentially sterile and passive consumers in the present (Miller and Rose, 1997). Surfers now rely on the most precise surf forecasts to decide when and where to go surfing, they have to have the latest wetsuit technology and the most state-of-the-art surfboards (some with embedded LED lights for night surfing), and they are constantly consuming new technologies for recording themselves – including surfing with waterproof cameras in their mouths, mounted to their surfboards, or hovering over their heads with drones. The end goal is now more efficient surfing (knowing when and where to go), meaning, not wasting time at the beach if the surf is not running, more waves in more places, having more and better images of surfing, to share with more people and faster over the Internet. The fate of surf culture is thus determined by the technologies created and consumed. This was not always the case, it is indicative of the technological era.

Technological determinism can be a dangerous concept. As evidence, Ted Kaczynski borrowed heavily from Jaques Ellul, one of the first to outline this idea, in his manifesto, titled 'Industrial Society and Its Future'. This manifesto outlined his rationale for committing terrorist attacks to begin a worldwide revolution against the dehumanizing effects of the 'industrial-technological system' (*Washington Post*, 1997). This is obviously an extreme interpretation, but if humans interpret their roll in a technological society as ineffectual and at the whim of out-of-control technological forces, then drastic/desperate measures make sense. This sentiment was also perhaps most popularly depicted in the film *Fight Club* – embodied in the character Tyler Durden's penchant for the destruction of artefacts symbolic of modern materialism (Lizardo, 2007). Like the Luddites who smashed machines to bring attention to the alienating impacts being brought on at the dawn of the Industrial Revolution, those who feel powerless against technique tend to lash out at the artefacts it produces in self-defence or in preservation of humanity. The Luddites are often misunderstood as a group of people who hated technological advance in and of itself, but in reality they were responding to the way technological advances in production caused workers to be laid off and put greater demands on those who remained employed. The Luddites smashed textile machines because the technologies symbolized social change that they perceived negatively.

In this sense, there is a longstanding precedent for the contemporary surf Luddites who smash and steal real-time wave cameras in order to keep information about their local surf breaks unavailable to the wider public (Mach, 2014; Kilgannon, 2008). Understanding technological determinism is a critical starting point for discussing the deepening commitment to technological advance taking over surf culture. It is an entry point to a new debate, it is not the coda. What is often interpreted as technological determinism is what Nye (1998) calls technological momentum. This thesis upholds near-term tenets of determinism, in that past decisions regarding production and the artefacts produced do inhabit the milieu at any given time and cultures are locked into that reality. What is critical, however, is that humans can interpret the impacts of deepening technological commitments and begin to exert power over technique – human actors can influence the social, political, and economic system to dictate what is produced and for what ends.

Technology, environment, and society (TES) and the evolution of surf culture

A coevolving TES framework is a useful theoretical framework for situating a discussion about a potential shift from the technological era towards another epoch. Within this framework technology can and does alter social connectedness and ecosystems stability, but rather than seeing this as a one way influence, societies can react to changes by organizing and guiding technological advance in new directions (Norgaard, 1994). Although I argue that we are presently in the technological era of surf culture it would be naïve to presume that that is somehow the endpoint of some Hegelian linear historical trajectory. The surf eras that opened this chapter changed due to evolutionary forces from one into the next (Mach, 2014). The TES interrelations that characterized each era eventually broke down and made way for a new cultural Zeitgeist. Saint-Simon (1975), who helped spearhead, lived through, and wrote extensively on the French Revolution suggested that historical change is characterized by a constantly swinging pendulum between organic and critical epochs. Organic epochs seem determined because they carry on without much resistance, but eventually latent discontent with the prevailing order of an era (often coupled with external inputs) manifests into a critical period where constituents begin to fight for change and usher in a new paradigm. While the current technological era seems organic or stable and destined to follow in a predictable path, understanding the history of surfing, using the TES framework as a guide, reveals how historic changes occurred and this lends insights into to where future change might come from. This framework opens up space for asking whether or not the pendulum is swinging in surf culture from a technological epoch to a sustainability one.

In the fifth century, through some mysterious mental processes, indigenous ingenuity, environmental circumstances, and social collaboration, Hawaiians brought the surfboard into existence – certainly an element of material reality equally capable of not being. There was, therefore, technological progress and technology dating all the way back to the inception of surfing. What has changed, however, is the ability of society to control or steer technological progress in a direction towards normative goals for societal maintenance and ecological sustainability. Hawaiians set parameters for surfing directly into their Kapu system, or system of laws that ordered social behaviour. Surfing was a part of the culture, but it was also constrained and governed by laws geared towards ecological and societal maintenance. Cultural ecologists maintain that values, kinship, customs, rituals, and taboos are related to the maintenance of the society's interaction with its ecosystem (Norgaard, 1994: 41). For Hawaiian surfing, the conscious desire to steer human–ecosystem interactions translated into rules dictating how to make surfboards, what materials should be used, and which rituals should accompany surfboard manufacturing, and these tenets were all built directly into the Kapu system (Stratford, 2010).

Environmentally, surfboards were made from indigenous trees and cutting down a tree in Hawaiian culture required planting two in its stead (Kotler, 2006).

Surfboards were also markers of social class, meaning technological constraints had, in another respect, a societal function (Stratford, 2010). Just by looking at someone's surfboard at this time you could tell their skill level and caste. Commoners constructed boards from the koa tree and the styles and sizes changed as different levels of skill where achieved. Commoner boards ranged from six to 14 feet depending on talent level. The ruling class used an *olo* board that was between 14 and 24 feet in length. This board was made from the wood of the more buoyant and rare wili wili tree. The Kapu system dictated that a Kahuna (skilled craftsmen) would search for the right tree, sacrifice a kumu fish as an offering to the gods and stand guard over the specimen overnight under prayer (Clark, 2011; Stratford, 2010). The type of coral used to shape the wood and the oils used to finish the board were also dictated and further rituals of dedication were practised before the board could be taken into the water. The inception of the surfboard changed Hawaiian social life and in turn, the social milieu influenced technological change, or the lack thereof. Hawaiian law directly stated how and with which materials commoners and elites must construct their surfboards. Tinkering with modifications to the ancient surfboard or using other materials was not considered desirable technological progress, but unlawful.

Swells, surfboards, and the activity they enabled were sacred. Hawaiians built alters to pray for surf and flew kites to signal that the waves were breaking to communities farming inland (Westwick and Neushul, 2013). High stakes were placed on surfing events and contests were used as a way to arbitrate disputes amongst clans and men and women courted one another in the surf because they were separated in many other spheres of life (Westwick and Neushul, 2013). Hawaiian life was built around surfing and surfing had sanctioned social functions, but surfing was also made sense of and controlled in furtherance of societal and environmental demands. The end of this era can be attributed to changing TES interactions brought on by the intrusion of foreigners after Cook arrived in Hawai'i at the end of the eighteenth century. Europeans brought new technologies and ideas about how to govern society, they also brought diseases for which the indigenous population had not developed immunity. As the population died off so too did the role of surfing in the society. Some surfers remained, but the activity was relegated to the margins, rather than a cultural centrepiece (Walker, 2011). When the Kapu system crumbled, so too did the arrangements that governed technological progress in surfing.

Though there were many surfing eras following Western adoption, the seeds of the technological era were planted as soon as North Americans, Australians, and Europeans began dabbling in the activity. This transition really took off following the Second World War when war torn nations tried to sort out an existence beyond constant strife and conflict, largely turning to leisure as a deserved respite from depression and war and surfing as a new activity around which to build identity and a sense of community (Bell, 1976; Ford and Brown, 2006). The post-war consumerist society had no protections against technological advance; in fact, society was becoming fashioned around the idea of unencumbered technological progress as a means to perpetual economic growth. Surfing was not immune to this transition, rather, surf culture, fuelled by representations of surfing in popular media and in advertising campaigns, began to lure more and more participants and the surf industry was built around this desire to participate – and technology the means for surmounting barriers to economic growth and participation. To get surfboards in the hands of more people in more places required technological advance to shrink and lighten boards so that regular folks living far away from the coast could store their boards and transport them to the surf. Cold water was also a limitation that needed to be surmounted, so industrious surfers began to adapt ways to create and distribute wetsuits to facilitate surfing in more and more places. The consumer capitalist system drove this initial phase of technological advance to help make surf products into profitable consumer commodities for an audience outside the original coastal enclaves. As the global surf population now swells to more than 30 million, technological advance continues to push the frontiers, making surfing accessible to more and more consumers.

One pertinent question associated with this chapter is what is driving technological advance and to what ends. Are we trapped in a web of the constant search for the most efficient and one best way to predict surfing conditions, get to far off waves, and ride them? Is technology constantly and creatively destroying that which came before as an end in and of itself or for some other purpose? How does the technological wedge influence human–nature interactions? This chapter calls for a better understanding of the ways in which the changing technological parameters facilitate and/or detract from efforts to usher in more sustainable surf practices. In order to move towards answering these questions and to characterize the technological surf era, this chapter will isolate and discuss the four technological dimensions (physical, climatology, Internet communication technologies (ICTs), and artificial surfing) impacting surf tourism (Figure 3.1). Aggregating these four dimensions will reveal how central technological change is to understanding how surfing and surf tourism evolves, how surfers interact with resource units, and the governance of common-pool wave resources.

The four technological dimensions

The physical dimension

Equipment

This section is specifically dedicated to physical surfing artefacts such as the foam/fibreglass, stand up paddle (SUP), and soft-top surfboards, as well as the leash and the wetsuit to illustrate how technological advance alters society and environment as these advances facilitate greater participation in surfing and open up new areas participating in the activity. This entails mentioning how the surf industry appropriated technological advances from other sectors to be incorporated in designing surf artefacts in order to reveal how deep-rooted and nuanced the commitment to pushing the technological frontiers in surfing has become to

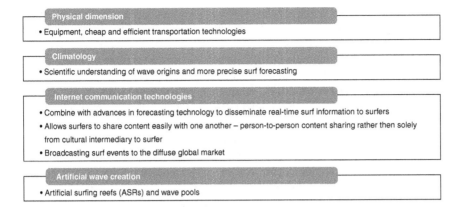

Figure 3.1 The four technological dimensions characterizing the technological era.

the activity. The industry borrows and adapts technologies from other sectors, as well as invests in divisions within companies to develop and advance their own technologies in house. The goal is to accelerate consumption through technological advances that facilitate more: more waves, more turns per wave, more barrels, and more time on the wave, as well as equipment (i.e. boards and wetsuits) that is 'best' matched to the conditions.

The path to incessant physical equipment advance in surfing began when many surfboard shapers in the Cold War era began to incorporate wartime defence technologies into surfboard design (Westwick and Neushul, 2013). Prior to this, surfers either knew how to craft their own boards out of local wooden resources or knew someone reasonably well who had the knowhow (Rensin, 2008). The foam/fibreglass surfboard changed surfing entirely. Foam/fibreglass surfboards introduced synthetic materials (rather than materials forged from the earth's innate regenerative capacity), made surfboards smaller and lighter, more buoyant, water tight, and also made boards more affordable and easier to mass produce. In the 1960s and 1970s surfboard manufacturing warehouses were springing up in Hawai'i and California to try to keep pace with the demand for surfboards. In the process small-scale craft manufacturing was being supplanted by mega manufacturing processes in an attempt to reap the benefits associated with mechanization and economies of scale.

Socially, this process removed the interconnectivity between surfer and the craftsman who designed and manufactured surfboards – a defining cultural characteristic of the pre-technology era. It also changed the nature of the way humans connect with waves by moving the activity away from using indigenous trees as surfboard feedstock and instead placing the focus in scientific board engineering with synthetic materials to improve riding efficiency and manoeuvrability. Environmentally, the toxic chemicals and petroleum by-products utilized in the manufacturing process have life-cycle impacts that include human health complications associated with factory production and the contribution to the waste stream both in production and after the life of the surfboard (Laderman, 2014). The most high-profile example involves Clark Foam, a company that monopolized the PU blank (base material for the majority of surfboard production) business and once controlled 70 per cent of the market, which was exposed for emitting more than 2,700 pounds of volatile organic compound into the California air during its annual production of nearly a half of a million blanks per year (*Surfer Magazine*, 2012). While this has led board blank manufactures in the US to become more environmentally conscious it also has pushed production to areas in low and middle income countries (LMICs) with fewer labour and environmental restrictions. This has had the dual impact of expanding the supply chain and deflecting rather than addressing the adverse implications of these processes. When coupled with the physical waste stream as surfboards reach the end of their useful lives, the mass production of boards to a wider audience has had severe environmental consequences. Especially considering in a recent poll of more than 2,300 surfers, 74 per cent reported owning two or more surfboards (Ponting, 2015). If there are 35 million surfers worldwide (O'Brien and Eddie, 2013), and roughly 75 per cent of them have two or more surfboards that is a great deal of consumption, replacement, and disposing – especially considering the surf population grows between 12–15 per cent (Buckley, 2002a) a year and a surfer's typical introduction into surfing is purchasing a board that they will often get rid of relatively quickly as they develop more skill.

Technology has also become the means for overcoming barriers to entry into surf culture. Manufactures continually press to incrementally make boards that are safer for introductory use (soft-top surfboards) and easier to paddle into waves (SUPs). Leonard (2013) reports that Costco has sold more than 100,000 of its Wavestorm soft-top model in its first eight years of distribution. This safe and easy-to-ride board, retailing at a big box store for around US$100, has literally changed the way that people become introduced to surfing. Rather than buying a used surfboard or buying a board designed by a shaper for between US$500 and US$1,000, consumers who want to try out surfing purchase these soft boards to reduce the risk of injury and the cost associated with owning a board. New surfers can both enjoy riding the cheap soft-top and get the hang of it and move on to other boards, or they give surfing a try and then get rid of the board if they don't enjoy surfing. These boards populate surf schools and surf breaks all over the world and have drastically increased access to the activity and consequently the competition in the surf for waves.

Stand-up paddleboarding has been named the outdoor activity with the highest percentage of first-time participants in 2013 with 863,520, up 56 per cent from the previous year (Lunan, 2013). SUPs have become popular because they allow surfers to stay standing at all times and use an oar to propel themselves into the surf in a way that is much easier than paddling in laying prone with one's arms and then needing to pop-up into a standing position. These new technologies have led to exponential growth in the surf population and also conflicts over resource use as novices on soft-tops compete with people who have spent decades developing surfing skills with more conventional technologies and fast paddling SUPers compete with regular surfers (Fletcher *et al.*, 2011). If surfing has any misanthropic undercurrents, this transition to a surf culture populated by mass produced/easy-to-use boards has brought this sentiment to the fore and also increased the waste stream associated with surfing as the demand requires more material throughput and energy consumption.

Lastly, the creation of the leash and the rubber/neoprene wetsuits, which were also said to have benefited from naval research and development, have also expanded access to more people in more places (Moore, 2010; Westwick and Neushul, 2013). The leash makes it easier and safer to surf because rather than losing the board all the way to the beach after every fall (requiring strong swimming skills) the board is right there after a fall to act as a flotation device. Once the wetsuit was adapted to surfing, this gear facilitated the adoption of surfing to colder climates further north in California from the epicentre between Los Angeles and San Diego, such as Santa Cruz and San Francisco, but also made surfing in Southern California's chilly water more comfortable and desirable. This technology makes serious year-round surfing possible in many other regions with large surf populations such as the east coast of the US, Australia, France, the UK, Japan, Peru, Canada, and Chile, to name a few. This development has also allowed surfing to spread to some of the coldest and most peculiar corners of the world such as Antarctica and Iceland (Moore, 2010). Wetsuits are now typically made of rubber and neoprene which require petroleum by-products in the manufacturing process and have a long decomposition half-life. These suits are also made with varying degrees of thickness so that surfers can be efficient with their use, meaning they can wear thinner neoprene in warmer conditions and thicker suits for colder climates. This ensures that there is an incentive to purchase an ensemble of suits and also to keep up with the latest updates in material science and flexibility. The wetsuit cannot be underestimated as a military technology adapted to surfing, which facilitated the expansion of the surf world to once unimaginable places, and manufactures continue to make these more efficient at keeping surfers warm with less volume and restriction.

Transportation technology

In surf literature, surf tourism is perhaps the most studied attribute for its contributions to facilitating social and environmental challenges in low and middle income countries (LMICs). Surf academics and journalists have gone as far as to contextualize surf tourism to LMICs as a form of colonialism whereby well-capitalized foreign entrepreneurs syphon the economic benefits from surf tourism, while placing the environmental (damaging coral reefs, over-extraction of food and building material resources to support tourism, the failure to deal properly with solid waste and effluent, removal of mangroves and other ecosystems for development, etc.) and social burdens (gentrification, loss of local cultural identity, tenuous employment, prostitution, drug use and sales) on host environments and communities (Ponting *et al.*, 2005; Ponting, 2008; Barilotti, 2002; Buckley, 2002b; Hughes-Dit-Ciles, 2009; Larson, 2002). No technology has facilitated this process more so than advances in air transport, which has made relatively affordable air travel to even the most remote coastal areas a common practice for many surfers (Barbieri and Sotomayor, 2013).

Furthermore, en route towards the 'one best way' (most efficient) to facilitate surf travel, surf tourism providers continue to integrate vertically their operations to make travel from the air to coastal surf destinations cheaper and easier. This removes barriers and helps surf trips to be hyper-specialized on surfing and con-sumed at even faster rates – get in, get out and get back. It is not uncommon for surfers to go from international flight, directly to automobile, to boat (some-times), to surf spot without stopping to do anything along the way besides use the bathroom. Iatarola (2011) found that these vertically integrated operations with their electronic advertising campaigns place local community run opera-tions at an economic disadvantage and jeopardize their ability to compete. All of this transport (to remote coastal surf destinations and between surf breaks), fur-thermore, has an unquantified carbon footprint, which potentially outstrips many other forms of touristic behaviour. While advances in air transport surely occurred for reasons outside of facilitating easy surf transportation, entrepren-eurs have leveraged these advances and continue to push for ways to facilitate faster and easier transportation to surf communities. This accelerates the spread of surf tourism and growth of remote coastal communities and exacerbates the negative social and environmental impacts above as developers seek to meet demand in an ad hoc and unplanned manner (Ponting, 2008).

Image capturing technologies

Participants are often lured to surfing after viewing images on surf websites, films, magazines, and advertisements. As long as there have been cameras surfers have been trying to capture themselves surfing. A defining element of the technological era is the personal waterproof camera that allows surfers to record themselves riding waves from different angles and points of view. Another emerging trend is filming surfing using drones flying overhead to get budget birds-eye-view shots of surfers. The implications of these technologies are essen-tially three-fold. The first being the reoccurring perceived obsolescence theme that these recording devices are constantly being improved and made more effi-cient, higher resolution, and user friendly than former models. The GoPro Hero 3 obviates the Hero 2 ad infinitum. This ramps up production and leads to more technological devices – which are rife with toxic chips, metals, dyes, and plastics – ending up with the waste stream. Anecdotally, we also know that these devices are often lost during surf sessions, also requiring new purchases. Second, these images are easily spread over social media (GoPro videos have received half a billion views in the last five years), which encourages more entrants into surf culture as people want to be able to do what their friends are doing and thus ushering in more competition for finite wave resources. Lastly, there are social implications associated with the primacy of capturing surfing manoeuvers using such devices. Paumgarten (2014) suggests that for the 'record everything' generation of adventure athletes in the technological era the 'agony of missing the shot trumps the joy of the experience worth shooting'. The surfer becomes a 'filmmaker, a brand, a vessel for the creation of content' rather than just a human being enjoying interacting with nature. Even if a surfer has an amazing time riding waves, if the film or picture does not pass muster compared to other shots out there, the entire experience is discounted. Surfers also tend to accept greater levels of risk in order to 'keep up with the Joneses' which can lead to bodily harm (Paumgarten, 2014). In short, the experience is dictated by the quality of the images, capturing them, and sharing them, which may reduce the physiological benefits associated with interacting directly with nature for its own sake.

Physical dimension conclusion

Advances in surfing equipment and image-capturing devices, as well as the accelerated pace at which they are developed and consumed, have life-cycle environmental and social consequences. Selling more and more surfboards, wetsuits, leashes, SUPs, soft-top boards, and cameras exponentially contributes to the waste stream helping to clog environmental pollution sinks during the manufacturing and disposal processes. There are processes at work which push the frontiers of the physical dimension of surfing, meaning technique continues to push industry not only to eliminate barriers to entry though technological advance, but also to trump that which came before with efficiency improvements. Surfers are quick to adopt the latest and best for the promise of more waves in more places – and more often than not, to record these experiences with greater clarity, ease, and from different points of view. When equipment is made from synthetic and often toxic materials and the activity of surfing becomes dominated by the requirement to record the experience, surfing becomes less about communing with nature and more about increasing wave counts and quality of the images captured.

Transportation technologies are also critical to the spread of surfing and surf tourism, which also has global environmental costs (e.g. carbon emissions) and localized environmental and social costs. Personal images flooding the Internet coupled with advances in beginner-friendly equipment have increased the demand for surf travel and the industry has been quick to bundle transportation and accommodation services to make the journey easy, focused on surfing, and consumed quickly. The three-fold physical dimension of the surf technology discussion illustrates how technological progress continues to march on unabated to accelerate manufacturing, distribution, purchasing, use, and discarding and also to mitigate barriers to entry into the surf culture. A growing surf population that consumes equipment, spreads images, and desires surf travel contributes to accelerating social and environmental challenges associated with consumption and competing for resources.

Climatology

Surfable waves were once mysteriously divine. They were something that Hawaiians literally prayed for during the foundational era. In the 1950s, surfers in California used to ask surfboard shapers when waves were coming as if somehow they could just sense their arrival because they could shape boards. During this time, surfers and shapers made the best guesses they could by observing prevailing winds and experiential trends, but their ways were more in-line with the occult, than mathematically precise (Rensin, 2008). Getting waves in the pre-technological era required dedication and close proximity to the beach. Having limited knowledge of when the waves were coming meant that you had to be there waiting and or constantly checking, which instilled a greater level of connectivity with nature.

Surfers surely have always desired better and more precise knowledge about future wave heights and tides, but the extent to which this information is now codified would have never been possible without the military discovering the need for and funding the installation of buoys and bathometric mapping all over the world. While surfers are perpetually looking for large surf, the military, after a series of botched amphibious strikes on coastal regions in the First World War (most notably the Dardanelles campaign) decided they needed a way to avoid big surf and low tides during Second World War invasions by sea (Westwick and Neushul, 2013). In order to establish the criteria for favourable sea conditions for invasions, oceanographic researchers, spearheaded by Dr Walter Monk at the Scripps Institute of Oceanography, developed a method for connecting wave data at a particular beach to weather data thousands of miles across the ocean (Galbraith, 2015; Westwick and Neushul, 2013). They then connected this weather and swell information with the way waves broke on a particular beach by factoring in the bottom contours and topography to scientifically isolate optimal times to deploy troops for land invasions by sea.

Building on Dr Monk's research, industrious surfers in the post-war period immediately began putting this information to use in order to stake out surf sites and optimal times to stage their own strategic strikes. This technology was eventually appropriated and turned into lucrative surf forecasting businesses. In 1985 Sean Collins, using physical charts and weather forecasts began a call-in service where surfers in California could pay per call in order to get surf forecasts and hints about the best places to go surfing based on this information. In the early 1990s this was turned into an Internet-based site and it is now a global site with information and videos available for thousands of surf breaks and a platform that is viewed by more than 1.7 million unique visitors per month (Surfline, 2015). There are also many other competitors in this space and the next section will discuss the implications of the availability of this information on the Internet in the technological era.

Appropriating advances in climatology to aggregate weather, swell, and near-shore bathymetry information help surfers to find new areas for surfing, as well as changes the fundamental understanding of what waves are, where they come from, and how they could be measured. In this process, ocean waves become 'demystified' and objectified, as they become known and understood based on calculable objective qualities such as swell size, period, and direction and near-shore wind. 'The technical approach to life focuses exclusively on what is rational in terms of striving for the greatest efficiency and externalizing every-thing else' (Vanderburg, 2005: 429). The adoption of atmospheric science to surf forecasting is a major component of the technological era that attempts to remove the wonderment and unpredictability of nature and replace this with techniques for objectifying and predicting swells – making the natural processes seem like an orderly code that can be cracked with high science. Interaction with the natural world through surfing has never been more mediated by scientific data and this has implications for human–nature connectivity in that the norm-ative goal becomes about performance ratios and more waves rather than com-muning with nature or psycho-social well-being. Knowing when to go has implications that are rarely discussed or examined; surfers use the information because it is out there and contributes to convenience rather than for other ends.

Internet communication technologies (ICTs)

ICTs have revolutionized surfing and are a defining characteristic of the techno-logical era. This section will discuss the three critical facets of the way in which ICTs impact surf culture: facilitating access to surf data, forecasts, and cameras, enabling peer-to-peer information sharing, and providing a mode for directly reaching the global surf audience with contests and events. The first facet of ICTs ties back into the last section of this chapter and has to do with how the demystification of surfing waves through deciphering objective data about current and future surf conditions is made publically available and used by surfers. 'Surfing the Web' has long been used as a phrase for describing aim-lessly clicking around the Internet with no real goal. The phrase reflects the widely held societal stereotype that surfers are goalless wanderers who lack dir-ection and purpose. The irony is that while many still get lost mindlessly click-ing around on the Web, the surfer has used available scientific and technological information about surf conditions to become strategic and precise in pursuit of quality surfing conditions. Technological era surfers rarely wander around phys-ically hoping to stumble upon good surf, they research the conditions, watch live surf cameras on their computers and mobile devices, connect seamlessly with one another, and plan surf trips based on information available for free to anyone on the Internet. This also allows surfers to be more productive in their careers as they can plan to go surfing when conditions will be good, rather than wasting time physically searching the beaches, adding another layer to the efficiency and productivity facilitated by ICT use, whereby surfers armed with technological knowhow can be productive citizens and not wave-chasing vagabonds.

The argument is that 'in an information society, a great deal of experience is displaced by information' (Vanderburg, 2005). Desk-side indoor surf checks remove a layer of connectivity with the coast and the sea. It goes even farther now that surfers can check surf reports and watch videos on their smart phones or get texts alerts when quality waves are predicted to be on the way. Surfers are surrounded by technique as technology surveils the oceans. Interaction with the natural world is now mediated by surf information and cameras available through ICT platforms and for a 'technique based connectedness, more is always better.... The only real human choice is an implicit one, namely, that performance values (in actual facts, technical ratios) take precedence over any cultural values' (Vanderburg, 2005: 429). The decision whether or not to surf and where to go surfing has become a rational one based on information rather than experience.

Evidence of a technological cocoon encircling the technological era surfer is available in the results of an ongoing survey of the global surf community. While presenting preliminary data from this survey at the Sustainable Stoke conference at San Diego State University, Ponting (2015) suggested that of the nearly 2,400 respondents, 34 per cent say that they look at surf forecasts on the Internet multiple times each day and 24 per cent say that they look at least once a day. Furthermore, 68 per cent of respondents agree or strongly agree that surf forecasting websites influence whether or not they go surfing each day and 67 per cent agree or strongly agree that these websites influence where they go surfing each day (Ponting, 2015). This data has not been formally published yet and is limited because the respondents were solicited from an online survey (may skew responses about Internet use) and the sampling was convenience-based. This data can, however, be interpreted in a number of different ways, but it is safe to say that the decision to surf and where to surf has become mediated by technology for many surfers. Surfers are not going to the beach, looking at or sensing conditions, and sharing information directly with friends as often as they are clicking around on their computer screens to see what is going on in the water. Interestingly, only around half of the surfers polled said that surf cameras and surf forecasts contributed to their enjoyment of surfing. So if not to facilitate greater levels of enjoyment, why use this technology?

Other than using ICTs to access surf data and cameras, surfers also use this platform to connect directly with one another. Prior to the ICT era in surfing, a great deal of information disseminated to surfers was created by cultural intermediaries (Bourdieu, 1984) who essentially had a monopoly on representing surfing around the world to the global surf population. These surf magazines and videos tended to show fetishized content (surf porn) which depicted remote coastal communities around the world as virgin wave utopias waiting to be discovered and exploited and they almost always left out the realities of the impoverished communities living around the waves (Ponting, 2008, 2009; Barilotti, 2002). These images are often said to have inspired a form of surf travel where surf tourists populate a tourist bubble (Carrier and Macleod, 2005) where they are sheltered from the same on-ground realities left out of the imagery that inspired them to travel (Ponting and McDonald, 2013). In the ICT era, it can be argued that surfers often seek to capture images of themselves surfing in far-off places in a way that mimics surf porn and spreading these images on social media platforms further encourages bubble-style travel which is offered as a packaged commodity from tech-era surf tourism providers (Mach, 2014). In short, peer-to-peer image and information sharing not only encourages new entry into surf culture, but also perpetuates a mode of disembodied tourist behaviour – both with social and environmental implications. Socially the activity becomes dominated by image capture and dissemination while environmentally, more surfers require more gear and material throughput surf tourism means straining resources and polluting remote coastal communities.

The last layer of the ICT dimension is the recent phenomenon of broadcasting high-profile surf competitions directly on Internet platforms. These increase exposure to surfing and creates a social media buzz around competitions. According to Minsberg and Corasanti (2015) 6.2 million viewers watched the World Surf League (WSL) pipe master's contest in Hawai'i via a YouTube Internet feed, and this is more viewers than is typical of the Stanley Cup finals in professional hockey. This reveals the growing popularity of the sport and the way in which ICTs help to spread surf viewership around the world in a way traditional television networks have been unsuccessful at in the past. Other contests airing online have brought, or will bring countries like Tahiti, Indonesia, and Mexico onto the radar of potential surf tourists, and this potentially has implications for surf tourism flows internationally. Though it has not been studied in surf tourism directly, anecdotally, based on studies in other fields, we can assume that exposure to surfing breaks in places where these contests are happening increase demand for surf tourism services in the short to medium run.

Artificial surfing

The most directly representative trend of the growing commitment to technological advance is the march to create high-quality artificial waves both at sea and on land. This process is about making surfing waves directly controlled and predictable though advances in engineering. Philosophically, the current obsession with creating artificial waves that are just like the real thing – only controllable, devoid of ominous sea creatures, and more consistent – suggests that nature is unable to produce the setting surfers desire efficiently enough and that mounting crowds and long, flat spells can become nostalgic through a commitment to advancing artificial wave technologies. This begs the question, can technology replace nature in surfing? Is the ride all that matters and connecting with nature a mere afterthought?

While there are now many surf pools that are a part of water theme parks all over the country, the quality does not satisfy intermediate to advanced surfers. The quest is now underway to establish the one best way to create a quality standalone surfing wave capable of drawing surfers, rather than as just one attraction among many in a water park. The gold standard has become to create a high quality, repeatable surfing wave that is as cost and energy efficient as possible and as close to a naturally occurring six-foot barrel that can be created (Surf Park Central, 2014). It is beyond the scope of this chapter to outline the entire market space, but at this stage there are more than a dozen companies competing to deliver the first operational surf pool with ocean like quality.

It remains to be seen how these surf pools will be received, but the commitment to bringing them to market is indicative of the technical era and increasing commitment to technological advance. In addition to surf pools, engineers have been working for the last decade to create artificial surf reefs (ASRs) on the ocean floor with the promise of drawing surf tourism to economically declining coastal communities, but also to have tangential benefits such as creating aquatic habitats for marine species and preventing coastal beach erosion (Rendle and Rodwell, 2014). The idea is that by placing permanent artificial reef structures at the bottom of the ocean, surfing waves will become more predictable, their refraction angles can be controlled, and the quality assured, thus drawing many surfers to communities in need of tourism revenue, as well as spreading the surf population away from crowded surf areas (Fletcher *et al.*, 2011; Rendle and Rodwell, 2014). ASRs have been installed in the US (though this one has since been removed due to design flaws), Australia, India, the UK (also currently inoperable), and New Zealand and proposed in Brazil and Portugal, but studies have not yet shown that the quality of these waves and their benefits have justified the costly investment. In short, the push to control nature and make ocean waves predictable has yet to significantly enhance the surfing experience in the intended ways. In the technological era, however, these challenges will be seen as surmountable with greater commitments to technological advance. The march to create artificial waves will continue and these may become defining characteristics of the technological surf era.

From the technological era to the sustainability era

What the previous section revealed is that commitments to technological advance in all four technological dimensions have created the conditions for increasing the surfing population and expanding the geographical area where surfing takes place, as well as making surf time more efficient though improved forecasting and the dissemination of that information. As the surf industry comes to rely more and more on technological advance to perpetuate their requisite economic growth, technological advance becomes a normative goal for its own sake and surf culture can be argued to be trapped in this paradigm. Companies not working on the next best thing, efficiency improvements, or artefacts that make surfing easier to do, are destined to become obsolete. Surfers tend to gobble up and apply technologies once they are brought to market, rarely questioning their contribution to personal enjoyment, well-being, or greater societal and environmental ideals, while becoming simultaneously rendered inert by the promises of what technological progress can bring in the future. It is a dizzying cycle in which surfers are consumers and observers, rather than change agents.

Technology, environment, and society are, however, co-evolutionary forces. Although surf culture can be contextualized as undergoing an organic technological era characterized by blind and uncritical acceptance of further commitments to technique, humans can exert the agency required to challenge the new normal and usher in changes that control technological advance in furtherance of normative social and environmental goals. There is evidence that this is happening. At a closer glance, many signs of this occurring are embryonic in the present tense, but have potential to become the characterizing elements of a future sustainability era.

First off, there have been conferences and gatherings geared towards bringing together surf academics, surf industry managers, surf development organizations, and concerned surfers to aggregate people and ideas for ushering in a sustainability era in surfing. The first illustrative example is the Global Wave Conference that was held in October 2015, which partially took place at the Houses of Parliament in Westminster, London and this setting is indicative of the growing political importance associated with surf-related topics. The theme of this conference was to promote global efforts and innovation to protect waves, oceans, beaches, and marine wildlife. The agenda included solutions for protecting surf habitats, innovations in sustainability in the surf industry, and lowering the impacts of surf tourism (GWC, 2015). There was also a Surf+SocialGood conference hosted in Bali in 2015, which was an action-focused conference for 'using the unifying power of surfing to connect and bring about positive change' (Surf+SocialGood, 2015). Furthermore, the Center for Surf Research at San Diego State University strives to be a resource for sustainable surf tourism researchers, tourism operators, communities, governments, and surfers globally. This entity has published many journal articles and educates undergraduates (CSR, 2015). Also, in partnership with Plymouth Sustainability and Surfing Research Group the Center co-edited the book *Sustainable Stoke: Transitions to Sustainability in the Surfing World* (Borne and Ponting, 2015). Later the same year an international conference was organized around the central themes of the book. Both the conference and the book bring together thought leaders and influential figures in the surf industry to share ideas and foster new normative ideals for surfing, as well as to aggregate these ideas into a new socio-environmental movement.

There is also evidence that the call to action for surfers to drive change in channelling technological advance towards promoting a transition into a sustainability era based on new normative ideals is being adopted by some of surfing's most influential figures. In a subculture as rife with hero worship as surfing, shifting a cultural paradigm requires support from the top professionals and respected surfers. Kelly Slater, the winningest and most popular surfer in the world has recently announced he will be establishing his own clothing company which he says he created

> to smash the formula. To lift the lid on the traditional supply chain and prove that you can actually produce great looking menswear in a sustainable way … the last two years have been a huge eye-opener for me. It's clear now just how challenging it is for any brand to put sustainability at the forefront of their business and I'm proud that we're one of the few taking the lead.
>
> (Borne, 2015)

Rob Machado has his own environmental foundation and also encourages and supports sustainable surfboard construction and Dave Rastovich, operating outside the pro-surf circuit, has been in films championing environmental causes and promotes sustainable living, of which surfing with environmentally sensitive equipment is a part. While Rastovich admits that change is slow, he suggests that now that influential surfers are demanding products where labour conditions and environmental impacts are considerations, the industry is beginning to make strides (*Surfer Magazine*, 2015b).

The proliferation of surf themed non-profit organizations also signals an appropriation of technological advance (using ICTs to bring awareness to their causes) while at the same time working towards encouraging the adoption of new ideals and ultimately a new era in surfing characterized by putting human and environmental well-being at the forefront. Surf non-profit organizations focus on a transition towards sustainable equipment and clothing production as well as on limiting pollution in waterways and encouraging sustainable surf travel practices. Sustainable Surf, which is one such example, is focused on:

1 developing the rise and availability of dramatically more sustainable surf-boards;
2 operating a unique recycling programme that helps turn waste Styrofoam packaging into new recycled-content surfboard cores;
3 turning professional surfing contests into a living showcase of engaging sustainable lifestyle choices;
4 inspiring individuals to adopt an ocean-friendly lifestyle as their own.

The founders of Sustainable Surf argue that surfers, when well informed, will make decisions that take environmental impact into consideration when they purchase products and that if surfers demand environmentally friends technologies, that other subcultures will follow in this path. The point is that these organizations symbolize latent discontent with the technological era status quo and are attempting to force a shift through information dissemination and guiding towards a new ideal in material production.

In what follows, examples will be presented of movements towards the sustainability era couched within a few of the technological dimensions which were used to organize the last section. This discussion is not meant to be exhaustive, but important for isolating emerging trends that, depending on social will, can either become widely adopted and usher in a new sustainability era, or remain novelties in an era dominated by constant technological advance for its own sake.

Shifting the physical dimension towards sustainable life-cycle production

While the majority of the surf equipment industry remains dedicated to the status quo through outsourcing production to areas with weak or non-existent labour and environmental regulations (Laderman, 2014), there are companies steering technological advance in the direction of sustainable materials, production, and disposal, as well as organizations working towards certifying environmentally sound production practices. While many of these initiatives are in the early stages of development, they are symbolic of socio-environmental movement driven by surfers to channel the direction of equipment production towards 'less damaging' practices. This is indicative of a normative shift away from technological era ideals in which efficiency and cost reduction are the critical factors in production decisions.

Sustainable Surf, for example, has created an Eco-board certification scheme that certifies surfboards which are manufactured with either a foam blank made with at least 25 per cent recycled foam (or biological content), resin made from at least 15 per cent bio-carbon content with low or no VOCs, or that the surfboard structure is made from sustainably sourced biological or renewable material. These boards receive a logo and have been used in pro tour events by many surfers including Kelly Slater. There are also companies such as Hess Surfboards, which uses sustainably harvested wood, recycled foam, and bio-resins to create 'eco-friendly surfboards as just one piece of a holistic approach to sustainable living' (*Surfer Magazine*, 2015a). While these boards cost roughly double that of standard PU models, there is often a six to eight month backlog on orders (signalling a demand despite added costs) and Hess is striving to create boards made entirely out of wood and sealed with plant oils. Other companies are experimenting with upcycling trashed/broken boards to create new foam blanks and others are using algae and other plant based resins. Large companies like Firewire surfboards also use bio-resins to encase their surfboards but also have shifted towards using packaging materials that are completely biodegradable rather than using Styrofoam and other materials that are rife with pollutants and take centuries to break down in landfills. Companies like Matuse and Patagonia are also making wetsuits that do not use synthetic rubbers or petroleum products but instead use plant based materials. At present, many of these developments are in the niche (early adopter) phase but their presence in the market signals that technological advance in the physical realm can be steered towards accomplishing social and environmental ends, rather than just be geared towards cost cutting and dissemination to the learn-to-surf market.

Luddism as an epistemology

Advances in climatology will continue as will the manner in which these advances are appropriated to better forecast surfing conditions in more places. Currently, Surfline.com has surf reports available for more than 3,600 surf breaks and hosts video streaming for more than 330 of those breaks (Surfline.com). Surf camera video quality will most likely improve and be deployed in more areas at an accelerating rate. The challenge to inserting human agency in the realm of the intersection between ICTs and wave forecasting is to resist allowing technologically driven data and information to determine activity. If surfers are fixated on efficiency and getting more waves per session utilizing the most advanced data and video technology, their behaviour will be ordered by technology and quantifiably measurable performance ratios (i.e. the number of waves per session) and they will lose an essential element of connectivity with nature. Surfers need to ask themselves whether or not it is important to be surprised by nature and to be truly in awe of the way that it enables surfing to occur, or if it is fine that waves are broken down into objective criteria and predicted with greater levels of accuracy with technological advance. Some suggest there are psychological and social benefits to appreciating the ocean's feel and its many moods and this can only be done through spending time in and around the ocean while it experiences many different sea states (Nichols, 2014). Just showing up on the beach when the waves are breaking may have implications for human–nature connectivity, rendering surfing more about the act of wave riding than an appreciation for the source of the waves which make the activity possible.

Adopting Luddism as an epistemology means to interrogate the use of these technologies in the present tense and deciding to deliberately go without certain technologies at times, or in extreme cases, lashing out against machines. There is already evidence of surfers breaking and stealing cameras to keep streaming video footage off of their local surf breaks in order to make it harder for web-surfers to 'know when to go' (Kilgannon, 2008). Though not confirmed, but also in this vein, powerful actors have been suggested to have lobbied surf web-sites to remove surf forecasts on their local surf breaks. These are overt symbols that humans are beginning to recognize the impacts that advanced technological precision in surf forecasting and surf camera imagery have on their lives and a desire to do something about it. While these are extreme measures, they demonstrate the recognition that humans understand the way the technology organizes societal behaviour (with negative consequence such as causing crowding) and this is worth lashing out against. Just as industrialization era Luddites were demonized and eventually discredited as backwards folks standing in the way of inevitable and socially beneficial 'progress', these surf luddites are being interpreted in a very similar fashion. There is little debate in surf culture as to whether their actions symbolize the need for interrogating the way advances in surf forecasting are implemented and used and how humans are impacted by greater commitments to technological efficiency. Who benefits from access to accurately forecasted surf information and what are these benefits?

This is not simply binary, surfers do not have to choose between smashing machines and fully adopting all technological capacity available. As a more reasonable approach, surfers can decide to disconnect more and make surf decisions based on normative metrics other than performance ratios and increasing the likelihood of traveling to an area for the sole reason of the promise of more waves for that reason alone. Winner (1977: 331–332) outlines an approach to epistemological Luddism which requires using the self as research into human–technology relationships:

[G]roups and individuals would for a time, self-consciously and through advance agreement, extricate themselves from selected techniques and apparatus. This, we can expect, would create experiences of 'withdrawal' much like those that occur when an addict kicks a powerful drug. These experiences must be observed carefully as prime data. The emerging 'needs', habits, or discomforts should be noticed and thoroughly analyzed. Upon this basis it should be possible to examine the structure of human relationships to the device in question. One may then ask whether those relations should be restored and what, if any new form those relationships should take. The participants would have a genuine (and altogether rare) opportunity to ponder and make choices about the place of that particular technology in their lives. Very fruitful experiments of this sort could now be conducted with many implements of our semiconscious technological existence...

Rather than smashing machines, adopting this epistemology is more about experimentation through removal and reflection. This is required to evaluate the force technology has on social evolution in the present tense and work towards consciously guiding technological change or individually using technology to further normative goals other than increasing quantities and efficiency. In short, this requires one to unplug, see how it feels, and then reorganize the influence one allows technology to have in one's life. This epistemology calls for using surf reports based on their contribution to social and psychological well-being, and not letting behaviour be determined by data both locally and in decisions to travel abroad on surf trips.

Surf travel in the technological era is dominated by the strategic strike, meaning surfers check surf reports online, decide to travel to locations when the waves are predicted to be good, and use surf tour operators to get them to the best surf (Reynolds and Hritz, 2012). This limits down time in the countries visited and also favours foreign tour operators who have the capacity to invest in having widely viewed Internet profiles above small local tourism operators (Iatarola, 2011). Though it is clear surfers travel first and foremost to surf, one potential avenue for adopting Luddism as an epistemology may be to try to rely less on technology in terms of surf reports and booking accommodation and surf tours and just fly somewhere and try to figure it out. This approach may require more time and entail less surfing, but perhaps it would have more local benefits and contribute to enhanced levels of enjoyment. The point is, it is time for experimentation and dialogue around these issues, or the culture is fated to remain within the technological era long into the future.

ICTs and surrounding gaze politics

Peer-to-peer information sharing though ICT, though it has the potential to rebroadcast fetishized content, has equal potential to enable the creation of a new discourse in surfing. Surfer-to-surfer communication enables the spread of dialogue centred on amassing popular public opinion around issues such as pur-chasing sustainable products, supporting surf development organizations (SDOs), travelling consciously and sensitively, and supporting environmental causes related to environmental issues important to surfers (such as clean oceans and limiting coastal development) (Mach, 2014). Giving a voice to the masses also allows surfers to engage in a watchdog function using ICTs to expose exploitative and environmentally insensitive and unjust business practices (Mach, 2014).

ICTs used in these ways become a conduit for 'surrounding gaze politics', which can influence social phenomena and result in changes in practices through gathering digital public opinion around certain issues and events (Du, 2011; Warren and Zeng, 2013; Yee, 2011). Most surf-related non-profit organ-izations could not exist without support garnered through ICTs. Many surf demonstrations, such as the Save Trestles from toll road intrusion movement and the campaign to protect Punta de Lobo, Chile from coastal development were also heavily reliant on ICTs to build public support and ultimately achieve their preservation goals. The proliferation of SDOs and surf-related issue campaigns, I argue, symbolize the appropriation of technological advance in furtherance of social goals and humanitarian/environmental initiatives in surf culture. ICTs can thus be used as a conduit for changing TES dynamics in favour of greater social and environmental ends. If this digital public opinion stays in the digital realm, however, there is the danger that change will remain an idea rather than an outcome. ICTs can be a medium for positive social and environmental change, only if surfers using the interwebs are compelled to embody actions such as joining protest movements, supporting non-profits, volunteering, and buying environmentally sensitive products. There is evid-ence this is happening.

Artificial surfing

At present, ASRs and surf pools are only really critiqued based on functional-ity. Their performance, technical difficulties, and costs have been criticized, but the philosophical implications of the march to establish the one best way for accomplishing these engineering feats are rarely interrogated. This is indicative of the technological era – the march continues, surfers will evaluate the technologies once they become artefacts, and then technique will pro-scribe the process for technological 'fixes' for the efficiency and quality issues that arise.

The types of questions that should be asked in an era when humans influ-ence technique rather than are influenced by it are: why do we 'need' ASRs and surf pools? What will their implications be for surf culture and the environment? How will they change the way humans interact with the natural world? Surfers need to reflect on what it means to use these artificial wave technologies and put them into context about what it means to be a surfer and why they surf. Is the environment and being immersed in the nearshore ecosystem an essential component of surfing? Or is it all about riding a wave, how many waves we ride, how long we ride each, and how many turns we do on the wave – however the wave was created? Will surfers learning in artificial pools alleviate crowding in the ocean or will it exacerbate these challenges as more people gain surfing acumen faster? In closing, these are all critical questions to pose and build healthy discourse around if these technologies are going to be somehow become a part of a new sustainability era or shunned for their detraction from this effort.

Conclusion

As far as Greek mythology goes, Prometheus stole fire from the Gods and from these stolen flames, all subsequent technologies and inventions were born that gave rise to civilization. For his crime, Prometheus was eternally chained and bound, which Winner (1977) suggests can be seen as a metaphor for how humanity is now bound by technique. Surfing was once the quintessential embodiment of human spiritual interconnectivity with nature. Boards were once wooden and surfers would glide on waves which came from a mysterious and divine source. The waves dictated when it was time to play and the absence of waves a signifier for encouraging work. Since the post-war period, the TES dynamics changed as surfing became situated in the Western economic paradigm, which began a process whereby technology has a great (and perhaps dominant) impact on social and environmental facets of surf culture. We are in the technological era and there are many associated implications that have been discussed in this chapter. The era can organically proceed or be challenged through a process of socially evaluating technology in the present tense and inserting agency in the direction of future technological change. This effort will require experimentation with epistemological Luddism and constant reflection. The first step is recognizing the technological milieu and beginning a debate around the associated social and environmental implications. There are signs this is happening, but the current initiatives taking on the goal of guiding technological advance in new normative directions will need to reach a critical mass to breakdown the core tenets of the technological era. It is up to every surfer to decide whether to tacitly accept the status quo and let the technology era proceed, or support movements in a new direction to usher in a new and more sustainable paradigm.

References

Barbieri, C., and Sotomayor, S. (2013). Surf travel behavior and destination preferences: An application of the serious leisure inventory and measure. *Tourism Management*, 35, 111–121.

Barilotti, S. (2002). Lost horizons: Surfer colonialism in the 21st Century. *The Surfer's Journal*, 3, 11.

Bell, D. (1976). *The Cultural Contradictions of Capitalism*. New York: Basic Books.

Borne, G. (2015). Kelly Slater supporting sustainable fashion. Retrieved 17 November 2015, from www.huffingtonpost.co.uk/gregory-borne/kelly-slater-supporting-sustainable-fashion_b_8555494.html.

Borne, G., and Ponting, J. (2015). *Sustainable Stoke: Transitions to Sustainability in the Surfing World.* Plymouth, UK: University of Plymouth Press.

Bourdieu, P. (1984). *Distinction: A Social Critique of the Judgment of Taste.* Cambridge, MA: Harvard University Press.

Buckley, R. C. (2002a). Surf tourism and sustainable development in Indo-Pacific islands: I. The industry and the islands. *Journal of Sustainable Tourism,* 10(5), 405–424.

Buckley, R. C. (2002b). Surf tourism and sustainable development in Indo-Pacific islands: II. Recreational capacity management and case study. *Journal of Sustainable Tourism,* 10(5), 425–442.

Carrier, J. G., and Macleod, D. V. L. (2005). Bursting the bubble: The socio-cultural context of ecotourism. *Journal of the Royal Anthropological Institute,* 11, 2.

Clark, J. R. K. (2011). *Hawaiian Surfing: Traditions from the Past.* Honolulu, HI: University of Hawai'i Press.

Csikszentmihalyi, M. (1990). *Flow: The Psychology of Optimal Experience.* New York: Harper & Row.

CSR. (2015). Center for Surf Research. Retrieved 16 November 2015, from http://centerforsurfresearch.org/.

Du, C. (2011). *Following is Power, Surrounding Gaze Changes China – On the Chinese Blogosphere.* Master's Thesis. Retrieved 10 January 2014, from http://arno.uvt.nl/show.cgi?fid=120739.

Ellul, J. (1964). *The Technological Society.* New York: Knopf.

Finney, B., and Houston, J. (1966). *Surfing: The Sport of Hawaiian Kings.* Rutland, VT: C. E Tuttle Co.

Fletcher, S., Bateman, P., and Emery, A. (2011). The governance of the Boscombe artificial surf reef, UK. *Land Use Policy.* 28, 395–401.

Flynn, J. P. (1987). Waves of semiosis: Surfing's iconic progression. *The American Journal of Semiotics,* 5(3 and 4), 397–418.

Flyvbjerg, B. (2001). *Making Social Science Matter: Why Social Inquiry Fails and How It Can Succeed Again.* Cambridge, UK: Cambridge University Press.

Ford, N. and Brown, D. (2006). *Surfing and Social Theory: Experience, Embodiment and Narrative of the Dream Glide.* Abingdon, UK: Routledge.

Galbraith, K. (2015, 24 August). Walter Munk: The 'Einstein of the Oceans'. *New York Times.* Retrieved 12 June 2015, from www.nytimes.com.

GWC. (2015). Conference report. Retrieved 16 November 2015, from http://globalwave-conference.org/conference-report/.

Hughes-Dit-Ciles, E. (2009). *The Sustainability of Surfing Tourism at Remote Destinations.* Unpublished Doctoral Thesis. Plymouth University, Plymouth, UK.

Iatarola, B. M. (2011). *Beyond the Waves: Economic and Cultural Effects of the Global Surf Industry in El Tunco, El Salvador.* Unpublished Master's Thesis. University of California at San Diego, La Jolla, California.

Irwin, J. (1973). Surfing: The Natural History of an Urban Scene. *Journal of Contemporary Ethnography,* 2(2), 131–160.

Kilgannon, C. (2008). Cameras show if surf is good, but surfers are getting in the way. *New York Times.* 27 January 2008.

Kotler, S. (2006). *West of Jesus: Surfing, Science, and the Origins of Belief.* New York: Bloomsbury.

Laderman, S. (2014). *Empire in Waves: A Political History of Surfing*. Berkeley, CA: University of California Press.

Larson, D. (2002). The making of a surf ghetto. *Surf Pulse*, September 2002. Retrieved 13 June 2013, from www.surfpulse.com/2002/08/the-making-of-a-surf-ghetto/.

Lawler, K. (2011). *The American Surfer: Radical Culture and Capitalism*. New York: Routledge.

Leonard, T. (2013). Wavestorm's surfboards reshaping industry. *The San Diego Union-Tribune*. 9 June 2013. Retrieved 9 June 2015, from www.utsandiego.com/news/2013/aug/31/board-wars/.

Lizardo, O. (2007). Fight Club, or the cultural contradictions of late capitalism. *Journal for Cultural Research*, 11(3), 221–243.

Lunan, C. (2013). OIA Outdoor Participation Report. In James Hartford (ed.), *SGB Weekly*, 1335. 9 September 2013.

Mach, L. (2014). *From the Endless Summer to the Surf Spring: Technology and Governance in Developing World Surf Tourism*. n.p.: ProQuest, UMI Dissertations Publishing.

Marcus, Ben. (2009). To Polynesia with love: The history of surfing from Captain Cook to the present. Retrieved 30 August 2013, from www.surfingforlife.com/history2.html.

Miller, P., and Rose, N. (1997). Mobilizing the consumer: Assembling the subject of consumption. *Theory, Culture and Society*, 14(1), 1–36.

Minsberg, T., and Corasanti, N. (2015). World Surf League takes web-first approach to drawing viewers. *New York Times*. 22 February 2015.

Moore, M. (2010). *Sweetness and Blood: How Surfing Spread from Hawaii and California to the Rest of the World with Some Unexpected Results*. New York: Rodale Books.

Nichols, W. (2014). *Blue Mind: The Surprising Science That Shows How Being Near, in, or under Water Can Make You Happier, Healthier, More Connected, and Better at What You Do*. New York: Little, Brown, and Company.

Noble, D. (1983). Technology's politics: Present tense technology. *Democracy*, 3(2), 8–24.

Norgaard, R. B. (1994). *Development Betrayed: The End of Progress and a Coevolutionary Revisioning of the Future*. London: Routledge.

Nye, D. E. (1998). *Consuming Power: A Social History of American Energies*. Cambridge, MA: MIT Press.

O'Brien, D., and Eddie, I. (2013). Benchmarking global best practice: Innovation and leadership in surf city tourism and industry development. Paper presented at the Global Surf Cities Conference. Kirra Community and Cultural Centre.

Ormrod, J. (2007). Surf rhetoric in American and British surfing magazines between 1965 and 1976. *Sport in History*, 27(1), 88–109.

Paumgarten, N. (2014). We are a camera: Experience and memory in the age of GoPro. *The New Yorker*. 22 September 2014.

Ponting, J. (2008). *Consuming Nirvana: An Exploration of Surfing Tourist Space*. Unpublished Doctoral Thesis, University of Technology – Sydney, Sydney.

Ponting, J. (2009). Projecting paradise: The surf media and the hermeneutic circle in surfing tourism. *Tourism Analysis*, 14(2), 175–185.

Ponting, J. (2015, September). Sustainable product demand in surfing. Working paper presented at the Sustainable Stoke Conference. San Diego State University.

Ponting, J., and McDonald, M. G. (2013). Performance, agency, and change in surfing tourist space. *Annals of Tourism Research*, 43, 415–434.

Ponting, J., McDonald, M., and Wearing, S. L. (2005) De-constructing wonderland: Surfing tourism in the Mentawai islands, Indonesia. *Society and Leisure*, 28(1), 141–162.

Rendle, E., and Rodwell, L. (2014). Artificial surf reefs: A preliminary assessment of the potential to enhance a coastal economy. *Marine Policy*, 45, 349–358.

Rensin, D. (2008). *All for a Few Perfect Waves: The Audacious Life and Legend of Rebel Surfer Miki Dora*. New York: Harper Entertainment.

Reynolds, Z., and Hritz, N. (2012). Surfing as adventure travel: Motivations and lifestyles. *Journal of Tourism Insights* 3(1), article 2.

Rosenberg, N. (1972). *Technology and American Economic Growth*. New York: Harper and Row.

Saint-Simon, H. (1975). *Henri Saint-Simon (1760–1825): Selected Writings on Science, Industry, and Social Organization*. New York: Holmes and Meier.

Scures, J. (1986). *The Social Order of the Surfing World*. Unpublished MA Thesis, University of Washington.

Stranger, M. (2011). *Surfing Life: Surface, Substructure and the Commodification of the Sublime*. Farnham, UK: Ashgate.

Stratford. (2010). Ancient Hawaiian Boards. Retrieved 12 June 2015, from www.clubofthewaves.com/surf-culture/ancient-hawaiian-board-building.php.

Surf Park Central. (2014, 1 February). Surf Park Business Model Viability – Jamie Meiselman – Surf Park Summit. Retrieved 12 June 2015, from www.youtube.com/watch?v=d_khcv6g0wA.

Surf+SocialGood. (2015). Surf+SocialGood – Splash. Retrieved 16 November 2015, from https://surfsocialgood.splashthat.com/.

Surfer Magazine. (2012). Clean up set: Clark Foam closes its doors. *Surfer Magazine*. Retrieved 9 June 2015, from www.surfermag.com/features/cleanupset_vol.47_3/#1HzfO5yafxo2gWdK.97.

Surfer Magazine. (2015a). The now: Danny Hess – shaper. *Surfer Magazine*. Retrieved 11 June 2015, from www.surfermag.com/the-now/danny-hess/#lt3IsJGdXZQSD0z5.97.

Surfer Magazine. (2015b). Unconventional wisdom: Dave Rastovich. *Surfer Magazine*, 9 March. Retrieved 19 November 2015, from www.surfermag.com/features/unconventional-wisdom-dave-rastovich/#4SyAgcym6WGdfJJQ.97.

Surfline. (2015). Media kit. Retrieved 21 November 2016, from http://i.cdn-surfline.com/advertising/mediakit/pdf/surfline_media_kit_2014.pdf.

Vanderburg, W. (2005). *Living in the Labyrinth of Technology*. Toronto, ON and Buffalo, NY: University of Toronto Press.

Walker, I. H. (2011). *Waves of Resistance: Surfing and History in Twentieth-century Hawai'i*. Honolulu, HI: University of Hawai'i Press.

Warren, R., and Zeng, Y. (2013). Social media and governance in China: Evolving dimensions of transparency, participation, and accountability. Paper presented at the 43rd Annual Meeting of the Urban Affairs Association, San Francisco, CA, 3–6 April 2013.

Washington Post. (1997). The Unabomber trail: The manifesto. *Washington Post*. Retrieved 12 June 2015, from www.washingtonpost.com/wpsrv/national/longterm/unabomber/manifesto.text.htm.

Weber, M. (1976). *The Protestant Ethic and the Spirit of Capitalism*. London: Allen & Unwin.

Westwick, P., and Neushul, P. (2013). *The World in the Curl: An Unconventional History of Surfing*. New York: Crown Publishing.

Winner, L. (1977). *Autonomous Technology: Technics-out-of-Control as a Theme in Political Thought*. Cambridge, MA: MIT Press.

Yee, A. (2011). China: Social media for social change. *Global Voices*. Retrieved 10 January 2014, from http://globalvoicesonline.org/2011/01/13/china-social-media-for-social-change/.

4 Towards more sustainable business practices in surf industry clusters

Anna Gerke

Introduction

This chapter deals with the local development of the French surf industry in the Aquitaine region and its impacts on the local economy. Tendencies towards more sustainable business practices are outlined using practical examples from the French surfing cluster. Sport leverages economic development directly through sport participation, spectators, tourism, and sales of sport equipment. Indirect impacts of sport are through infrastructure, hospitality, hotel and transport services related to sport (Dolles and Soderman, 2008; Preuss, 2005). Outdoor sports and economic development related to them tend to develop in places that provide favourable conditions to practise the respective sport. These observations have led to an important number of studies on localised sport industries, also referred to as sport industry clusters (Chetty, 2004; Gerke *et al.*, 2015; Glass and Hayward, 2001; Kellett and Russell, 2009; Parker and Beedell, 2010; Sarvan *et al.*, 2012; Shilbury, 2000).

Sport industry clusters are agglomerations of interconnected firms and associated institutions in a particular sport or group of related sports. Typical stakeholders of sport industry clusters include diverse sport equipment manufacturers (core equipment, systems, and accessories suppliers), services providers, media, designers, professional and amateur sport organisations, governing bodies, and education and research institutes (Gerke, 2014; Gerke *et al.*, 2015). This chapter explores factors that facilitated the development of the surf industry cluster in Aquitaine as well as organisational and industry-level strategies that have led to more sustainable business practices in the surf industry leveraging the cluster structure of the industry.

Research on corporate social responsibility distinguishes four dimensions: economic, legal, ethical, and discretionary (Babiak and Wolfe, 2009; Carroll, 1979, 1999). The economic dimension refers to the profit imperative of companies. The legal aspects relate to the duty of each company to obey the laws. Ethical motives encourage firms to act in line with societal expectations, while discretionary behaviour refers to activities that go beyond societal expectations. Sustainable business practices in this study are to be placed between the legal and discretionary dimension. They are defined as any behaviour that goes beyond what companies are legally required to do in order to improve environmental, societal, and economic impacts of their actions. I differentiate the actions of single companies (business level) and collective actions by a group of companies at industry level.

This chapter uses the case of the surfing industry in Aquitaine in the southwest of France. After presenting the case study, research methods and data collection instruments, it starts with an overview of historical and political aspects that have led to the development of a surf industry manufacturer association and later to a formalised surf industry cluster governing body. The chapter then outlines geographical and geo-economic aspects in the development of the surf industry cluster before exploring socio-economic and sport-related factors as levers of the surf industry cluster. Practices in the French surf industry cluster are analysed on firm level and industry level.

The surf industry cluster in Aquitaine

The French surf industry is primarily based between Hossegor at the northern end and Hendaye at the Spanish border in the south. It stretches over 60 km along the French Atlantic coast. The highest density of surf businesses and associated institutions can be found in the cities Biarritz, Hossegor, and Anglet. Figure 4.1 illustrates the geographical position of the French surf industry cluster.

Up to 400 firms have been identified with activities in the surf industry of which about half are members of a surf industry manufacturer association that embodies a surf cluster governing body (EuroSIMA, 2015c; Outdoor Sport Valley, 2014). The region Aquitaine estimates that about 3,500 people are directly employed in the boardsport industry (including winter, summer and urban boardsports) generating revenues amounting to €1.7 billion (Région Aquitaine, 2015).

The surf industry governing body had been initiated in 1999 by leading surf and boardsport brands with the intention to federate the actors of the European boardsport industry and to defend their interests. In 2005 this association – named EuroSIMA in accordance with the American sister association SIMA[1] – decided to include mountain sports and urban sports. Therefore EuroSIMA reframed the scope of the association as a European association of boardsport industrials but keeping the original name (EuroSIMA). Mountain sport was incorporated via a partnership with a mountain sport cluster in the French Alps (Outdoor Sports Valley). In 2008 EuroSIMA launched the EuroSIMA cluster in order to create a regional network between private and public actors in the Aquitaine region (EuroSIMA, 2015b).

This study is based on a single case-study approach in order to obtain deep insights into an under-researched topic (Eisenhardt, 1989; Yin, 2009). For this case study I approached the surf cluster governing body EuroSIMA in Aquitaine for collaboration purposes. Together with the cluster manager we identified a cluster member typology and key cluster members to interview for this study.

Figure 4.1 Geographical location of the French surf industry cluster located in the region Aquitaine in the southwest of France.

This was an inductive process that started with some initial types of cluster organisations that were identified from the information provided by the cluster governing body. There were, for example, major boardsport brands such as large-scale providers of generic surf equipment but also clothing and accessories as one type of cluster organisation. However, others emerged during the data collection process, for example equipment specialists that focus solely on one or a few often technologically demanding products or pieces of products; for example skateboarding shoes. Finally a typology of ten cluster members resulted from initial screening of the industry and first interviews that guided the remaining selection of interviewees for this study (Miles *et al.*, 2014). Table 4.1 shows the ten types of cluster organisations in the surfing cluster and how they fit with generic sport cluster organisations (Gerke *et al.*, 2015).

Table 4.1 Typologies of cluster organisations in surf industry clusters

Generic sport cluster	*Surfing cluster*
For profit commercial organisations	
1 core equipment manufacturer	surf/boardsport brand
2 system supplier	surf equipment specialist
3 accessory supplier	surf accessories and clothing
4 services provider/consulting	surf services/consulting
5 media/communications	surf media/communications
6 designer/architect	surfboard designer/shaper
For profit and non-profit sport organisations	
7 professional sport organisation	professional sport organisation
8 amateur organisation	amateur organisation
Non-profit and public organisations	
9 education/research institution	education/research institution
10 governing body	governing body

The data collection comprises 26 in-depth interviews of which 24 were semi-structured following an interview guideline and two were explorative. Table 4.2 provides an extensive overview of the organisations interviewed. In some larger organisations more than one person was interviewed but in total 21 different organisations were interviewed. Of those interviewed organisations 19 had 100 per cent of their activities in the boardsports industry. Companies and related institutions had arrived or been set up in Aquitaine from the 1980s until the early years of the twenty-first century. Most of the interviewed organisations are based in Hossegor.

The interviews served as the main source of information but were complemented with observations and secondary data. I conducted three observations at three different kinds of events. The first was the annual meeting of the EuroSIMA cluster in Hossegor, which allowed me to get in touch with cluster members, conduct informal interviews, observe interactions between cluster members, and follow current themes in the surf industry, for example a market research study on the skateboarding industry. The second observation took place during the Quiksilver Pro France, the French tour stop of one of the most renowned international surfing competitions. This observation provided an overview of the role of professional sport in the surf industry cluster and the implication of surf firms and other organisations in professional sport. Finally, I visited the trade show 'Gliss Expo Seignosse', which gathers together local surf industry businesses for several days to exhibit new products to the public and to business partners (Gerke, 2014).

Table 4.2 List of interviews

No.	Type of cluster organisation	In France since	Location	Interviewees' position
Formal semi-structured interviews				
1	boardsport brand	1980s	Hossegor	Core Division Manager/CEO Europe
2	boardsport brand	1978	Hossegor	Marketing Director
3	boardsport brand	1978	Saint-Jean-de-Luz	Product Line Manager
4	boardsport brand	1978	Hossegor	Global Technical Product Director
5	boardsport brand	1999	Bayonne	European Sales Director
6	boardsport brand	1980s	Hossegor	Technical Division Manager
7	equipment specialist	2010	Bidart	Director
8	equipment specialist	2009	Anglet	Director/R&D Engineer
9	accessories and fashion	2004	Saint-Jean-de-Luz	Marketing Director
10	accessories and fashion	1997	Hossegor	Marketing Manager
11	accessories and fashion	2000	Hossegor	Marketing Manager
12	services/consulting	2002	Bidart	Associated Director
13	services/consulting	2004	Biarritz	Director
14	services/consulting	2003	Anglet	Associated Director
15	services/consulting	1978	Bayonne	General Manager
16	media/communications	2002	Biarritz	Editor
17	design/shaper	1981	Anglet	Director
18	professional sport	2000	Hossegor	Marketing Manager
19	amateur organisation	1967	Aquitaine	Surf Equipment Consultant
20	education/research	1985	Bidart	University Professor
21	governing body	N/A	Bayonne	Director for Development, Research, Innovation
22	governing body	N/A	Mont de Marsan	Director for Economic Development
23	governing body	2005	Bayonne	Cluster Manager
24	governing body	1999	Bayonne	Director
Explorative interviews				
25	governing body	2005	Bayonne	Cluster Manager
26	boardsport brand	1980s	Hossegor	Core Division Manager/CEO Europe

Location-specific factors in the development of the surf industry cluster in Aquitaine

Location-specific factors are key in the location decision of enterprises, especially when envisaging an internationalisation strategy. Location-specific factors can be exploited if the company owns or is able to internalise the therefore required firm-specific advantages. The foreign company needs to overcome the liability of foreignness by leveraging firm-specific advantages (Dunning, 1980; Rugman, 2010). In the following paragraphs historical and political factors and milestones that were relevant for the development of the French surf industry in Aquitaine are outlined based on information from our data collection. In the same manner geographical and geo-economic factors are discussed as well as socio-economic and sport-related factors. Outlining location-specific factors of the surfing cluster sets the background and context for the analysis of motives for sustainable actions, practices, and strategies.

Historical and political factors in the French surfing cluster

In order to analyse the historical and political aspects in the development of the surf industry cluster in Aquitaine in France, we outline the beginnings of surfing in France as well as the development of the first European surf industry federating and governing body that was established in Aquitaine. It was a film crew that brought the first surfboard to France in 1956 (Falchi, 2006). This first surfboard was left behind and represents the starting point of surfing in France. Very soon the foundations of the French surf industry were laid during the 1960s with the first small surf businesses set up by surfers to serve local surfers' needs concerning surf equipment. In the 1980s the growing North American and Australian surf companies arrived in France and settled in the Aquitaine region to conquer the European market. It was only in the beginning of the twenty-first century that French authorities recognised and acknowledged the economic contribution and potential of the surfing industry at the local and regional level (Falchi, 2006).

The institutionalisation and formalisation of this traditionally rather unorganised industry, when compared with the sister industry in North America, started in 2005 following the North American model of boardsport industry federation, SIMA (Surfing Industry Manufacturing Association) (Jarratt, 2010). The cluster governing body EuroSIMA was established under the initiative of major surf and boardsport companies and recognised as a local productive system by the French government (Falchi, 2006). In 2008 this federating initiative was reinforced with the foundation of the EuroSIMA cluster aiming at the creation and development of a tight industry network that would generate economies of scale, learning effects, and synergies (EuroSIMA, 2015b). This was a consequence of the rejection of an application to obtain the French national label of 'pôle de competitivité', which in France is a more prestigious and recognised label than the unprotected term 'cluster'. The rejection was to some extent a reset for the collective and unitary efforts of the surfing industry in Aquitaine but allowed them to reunite and establish the EuroSIMA cluster (Falchi, 2006).

While the EuroSIMA cluster was initially an industry initiative, the regional government participates in the funding of the surf industry cluster federating and governing body and its activities. About half of the cluster governing body's budget is funded by public subsidies from different institutional partners, while the other half comes from member contributions, donors, and private partners, as well as income from the cluster governing body's activities (EuroSIMA, 2015a). Next to financial contributions government agencies provide a variety of assistance to the cluster governing body and its members. This includes mentoring, assistance, support of innovation projects, providing market information, networking, and international development and marketing (Gerke, 2014).

The EuroSIMA cluster provides specific services to its members. These include five main activity areas such as economic observation, innovation and R&D, sustainable development, collective actions for artisanal entrepreneurs, and HR and communication projects. The economic observatory provides regularly market reports on relevant industry sectors. Innovation is fostered by accompanying enterprises during their innovation projects, by managing innovation projects, but also by hosting innovation competitions and awarding innovation prizes. The EuroSIMA cluster usually works with subcontractors and service providers that are specialist experts in the various niche fields, for example an expert in plastic material and natural rubber was used for a project about neoprene wetsuit recycling. In a project around sustainable development the EuroSIMA cluster studied together with the support of some cluster members the possibilities to replace plastic bags in the packaging of clothing. Synergies are achieved through the bundling of job announcements on EuroSIMA's website to provide a space where demand and supply of this specialised workforce can easily meet.

Geographical and geo-economic factors in surfing clusters

The role of geographical location conditions as well as those of firm location and the spatial proximity of firm locations are investigated in the following paragraphs. Before outlining these aspects of the surfing cluster, the importance of the surfing cluster in terms of sales turnover is illustrated. In 2005, at the time of the application for the label 'pôle de competitivité', the Aquitaine surfing industry had a turnover of €800 million. In addition, around €285 million turnover accounted for winter boardsports, skateboarding, and other items. These numbers constituted approximately one tenth of the global boardsport industry. In its 2014–2015 activity report EuroSIMA (2015a) estimated the European boardsport market amounting to €13 billion and representing 31 per cent of the worldwide market.

The development of agglomerations of proximate firms and associated institutions around surfing has already been observed in Australia in the Gold Coast and in the Victoria region around the town Torquay (Stewart *et al.*, 2008; Warren and Gibson, 2013). Cluster theory argues for the importance of spatial proximity between firms to achieve interorganisational dynamics and synergies (Becattini, 2002; Porter, 1998). In the case investigated here interviewees emphasise that it was and still is important to be in the cluster region (Aquitaine) for image reasons and to have access to a network of both professional surfers and surf professionals. With the arrival of surfing in Aquitaine, the region quickly became the European surf capital in terms of surf holidays but also in terms of surf equipment. Small local artisanal surf shops for surfboard shaping, repair, rental, or lessons were established along the Atlantic coast from the 1960s. Foreign surf brands, primarily from the United States and from Australia, had grown big and established their European headquarters in the region Aquitaine by the late 1980s. New boardsport brands in urban sports and accessories continued to appear in the early twenty-first century. While the original motives for these kinds of businesses sometimes varied in the initial phase, all companies cited the region of Aquitaine, with its famous surf destinations such as Biarritz and Hossegor, as 'the place to be' and as the 'Silicon Valley of surfing' if you wanted to be in the surf business.

The first reason for being located in Aquitaine was initially the proximity to good surf conditions during most of the year. The Atlantic coastline is well endowed with surf sports that are rich in waves of various size and type, suitable for both beginners and advanced surfers. These geographical conditions attracted surfers initially to come to the Aquitaine region. In consequence of an increased number of participants in surfing and hence an increased demand for surf equipment, surf businesses were established. A unique feature of these surf companies was the typically flat hierarchy and a participative management style with flexible work hours. These companies and work conditions not only allowed but encouraged employees to continue surfing and taking part in other boardsports even when occupying time consuming managerial positions.

Another reason for being located in Aquitaine is the easy access to other surf professionals, whether it is to find the right material supplier for surfboard shaping, specialised media or event agencies, surf retailers and schools, and most importantly to meet the professional surfers that tend to train in the Aquitaine region. The co-location of this variety of surf-related firms and associated institutions allows frequent formal and informal meetings and close ties between the actors in these organisations.

Socio-economic and sport-related factors in surfing clusters

The influence of social processes on economic activities as well as the role of sport-related factors in the surfing cluster are outlined in the following paragraphs. In terms of socio-economic factors the interviewed firms refer to the high mobility of staff amongst the cluster organisations within the cluster but not outside of it. This is a typical characteristic for industrial clusters and allows the development of interorganisational networks since staff that change the employer within the cluster tend to keep social informal links to the former employer or colleagues (Asheim, 2000; Bellandi, 2002).

It is a shared passion for the sport of surfing that underpins the social ties amongst employees of different organisations in the cluster. The joint practice of surfing while being sometimes employed by direct competitors creates social links between employees across different cluster organisations via the sport. Those links persist despite changes in corporate or industrial structures because they are based on sport and related values rather than on affiliation to a particular organisation. The surf industry is a dense interrelated network of people that put surfing first in life. This is the principal value people in the surf industry share. Surfers came to Aquitaine to realise a lifestyle that integrates regular surfing practice with work. The growing local demand for surf products and services made this possible. However, the internationalisation of the surfing market with mostly North American and Australian firms has opened a market of non-surfers that consume surf products because of the values and lifestyle transmitted by the surf brands. At the same time this new market has made the surf brands vulnerable towards other non-surf related labels that 'freeride' on the surf image without having any link or providing any contribution to the sport (e.g. Abercrombie & Fitch).

Professional sport and related professional sport events play an important role for the surf industry cluster and its member organisations. The surf brands sponsor athletes and sport events and therewith contribute directly to the development of surfing. Beyond the excellent surf conditions, surf events also increasingly influence sport tourism and therefore take an increasingly important role in the surf industry.

The surf cluster in Aquitaine consists of a variety of firms and associated organisations that tend to engage in sustainable practices on the business and industry levels. Having outlined the background and context of the French surf industry cluster in Aquitaine, the following sections outline some examples of tendencies towards sustainable business practices.

Firm-level and collective strategies towards more sustainable practices

Location-specific factors influenced the emergence and development of the surf industry cluster in Aquitaine. Building on these factors and the sport cluster approach, firm- and industry-level strategies in the surf industry cluster in Aquitaine towards more sustainable business practices are outlined (Gerke, 2014).

Firm-level sustainable practices in the surf industry cluster in Aquitaine

In order to distinguish different firm-level strategies towards more sustainable practices I look at different types of cluster organisations as suggested in the sport cluster concept (Gerke *et al.*, 2015). Core equipment manufacturers in the surf industry correspond to the big boardsport brands (e.g. Quiksilver) that originally focused on the design and manufacturing of surfboards. Nowadays these firms have outsourced most of the manufacturing process and concentrate mostly on design, R&D, and/or marketing activities. However, initiatives towards more sustainable manufacturing practices at the firm level are visible in small and medium-sized equipment specialist that engage primarily in material or design innovation (Gerke, 2014).

One example is a surfboard manufacturer who uses environmentally friendly materials and processes. The core of the surfboard is made of 100 per cent recycled expanded polystyrene (EPS) and the outer layer is stratified with organically sourced epoxy resin. The fabrication process of a surfboard is traditionally toxic and causes health problems for surfboard shapers. With the introduction of ecological raw materials the manufacturing process has also been improved.

While this company is surely acting in the economic dimension of corporate social responsibility, the company has gone beyond that by setting new environmental standards in the surfboard manufacturing industry (Babiak and Wolfe, 2009). Several innovation awards have been awarded to this company (for example by EuroSIMA). They were one of the first to participate in a government initiative that evaluates environmental characteristics of firms and their products and certifies the evaluated products with an eco-label.

Another example of sustainable practices at the firm level is a small accessories specialist who develops and manufactures surf wax made of environmentally friendly ingredients. Sustainable practices are undertaken in the product development by replacing petrochemical materials that are typically used in comparable existing products with sustainable ingredients. The company is a young start-up and dedicated to the use of organic materials even if those are more expensive than traditional products. The business is based on material innovation towards the use of more sustainable products. The founder explains his motivation for this business with respect to the environment. This statement indicates that the motives for responsible behaviour in the context of this start-up enterprise go beyond the economic dimension by taking the risk of educating the customer about organic products while alternatives at lower prices are available.

The motivation of these entrepreneurs to engage in sustainable business practices through the development of products made of organic or environmentally friendly sourced ingredients can be related to some location-specific factors. Surfing is closely linked to the environment and nature since the practice takes place on beaches, often embedded in stunning natural landscapes. This sensibility of surfers and surf businesses towards the environment can be derived from the fact of being dependent on the environmental conditions in order to continue the practice, and hence the business of surfing, but also through the development of values related to nature and environment (Scarfe *et al.*, 2009). Therefore the geographical and sport-related factors are most likely to explain sustainable behaviour that is evident at the firm level.

Collective sustainable practices in the surf industry cluster in Aquitaine

Traditionally a company would contract an external research laboratory to assist with the product development. However, to call upon external organisations in order to address strategic issues like innovation can take place in other formal or informal ways. The next paragraphs develop interorganisational cooperation and collaboration as strategies to act sustainably by outlining examples of sustainable business practices where three or more organisations of the same industry engage jointly.

A collective initiative towards more sustainable business practices was initiated by a handful of boardsport brands together with the surf industry cluster governing body EuroSIMA. The purpose of this collective project was to find solutions for the recycling of neoprene wetsuits. Some surf companies had individually started working on research on how to recycle neoprene. While this was very cost-intensive for single firms, some abandoned individual efforts, and some continued, eventually finding applications of recycled material in another product range or another industrial application. EuroSIMA had the ambition to provide general access for all cluster members to a solution that allows the reuse of recycled neoprene in an industrial application. EuroSIMA looked for an industrial partner in this project and found one who is able to cut the neoprene into small pieces and fragment them in a way that they can be reused for the fabrication of caoutchouc. The caoutchouc is then sold to an automotive supplier that reuses it for automotive pieces (for example the fabrication of windscreen wipers). EuroSIMA has led this research project and established the necessary eco-system so that other surf companies of the cluster can follow this recycling process. It is a collective knowledge that is made available to all cluster members. The cluster governing body's ambition here is to change standards in the surfing industry and to facilitate sustainable behaviour throughout the product life-cycle, which indicates the discrete dimension of corporate social responsible behaviour (Carroll, 1999). The difficulties cited were that in spite of the willingness to cooperate for this particular project, conflicts of interest and competitiveness hindered or diminished the advancement of the project at certain stages. For example some companies used the recycling of wetsuits in order to apply commercial benefits to buyers. This was not appreciated by other contributors of the project since not all companies were willing to provide price discounts on new wetsuits when old ones were recycled. A race for the most sustainable brand image and reputation in the surf industry equally diminished the gains from this collective project.

Another example of a collective project towards sustainable business practices that was led by the EuroSIMA cluster governing body aimed at finding alternative and more ecological solutions to the plastic bags (poly bags) that are typically used to package new products, especially clothing and accessories. Millions of these highly polluting bags are used year by year. EuroSIMA took the initiative at the industry cluster level to conduct a research project that aimed at finding organic and biodegradable solutions and to make them accessible to the cluster members. The research part of the project was contracted to a research laboratory specialised in eco-conception and accompanied by the EuroSIMA cluster. The laboratory analysed different possible materials for packaging that could be recycled. The result of the study is a recommendation for an alternative product to plastic bags that specifies the material properties, the thickness and also the supplier. The final solution is recyclable paper which is already made out of recycled paper. This example indicates the discretionary dimensions of corporate social responsible behaviour at the industry level. EuroSIMA initiated the project to change industry-level standards in the cluster.

There are also examples of sustainable practices based on collective initiatives of very small artisanal types of firms. In the surfing industry these are typically the surfboard shapers. Also in this case the EuroSIMA cluster took the initiative to develop more sustainable practices. In close cooperation with selected surfboard manufacturing companies and surfboard shapers they developed a workshop and working guide that integrates a maximum of respect for the environment and the health of shapers. This guide is made available to all cluster members and with some time delay also to non-members in order to encourage health and environmentally friendly surfboard manufacturing processes and procedures. The 'Eco Guide' is realised in a cooperative surfboard workshop that is co-financed by all participants.

Reflections and critiques on sustainable practices in the Aquitaine surfing cluster

The surf industry has gone through intense developments in terms of the size and scope of the industry. Being originally localised and small the surf community and industry has grown to a global scale. The surf market embodies today not only active participants in surfing but a growing mass market of fashion and lifestyle customers (EuroSIMA, 2009; Falchi, 2006). These changing industry and market conditions have made the atmosphere of the surfing industry more competitive while being originally a localised niche industry (Asheim, 2000; Marshall, 1920).

While many surf companies maintain a mind-set of local and within-industry competition, market trends are indicating the change towards globalised and cross-industry competition. These signs have resulted in new strategic approaches on the firm and industry levels in terms of how to run a business and what to focus on. Increasingly surf companies work in collective approaches. Environmental and social issues and impacts through business activities are more and more taken into consideration by decision makers in the surf industry (EuroSIMA, 2015a).

However, the collective efforts in the French surfing cluster in the Aquitaine region remain marginal and only on a collective and sometimes cooperative basis. The surf industry governing body EuroSIMA is a key intermediary in these collective and cooperative projects. The current cluster governing body model has functioned for more than a decade but is a risky model since it still depends largely on public funding. In order to become a sustainable business model companies need to take initiatives themselves to approach larger, costly, and risky topics in collaborative approaches. Topics related to sustainability and corporate social responsibility are typically these kinds of general topics that are more easily approached on an industry level than on an individual business level.

Acknowledgements

I would like to thank the EuroSIMA cluster, and more specifically the cluster members and the cluster manager, Christophe Seiller, for their interest and support in this study. Furthermore I acknowledge the advice and support of my PhD supervisors Michel Desbordes and Geoff Dickson during this research project.

Note

1 SIMA = Surf Industry Manufacturer Association.

Further reading

This research is based on some of the data that was collected for my PhD thesis. In this regard I would like to refer to the following publications:

Gerke, A. (2014). *The Relationship between Interorganisational Behaviour and Innovation within Sport Clusters.* Unpublished doctoral dissertation, Paris-Sud University, Orsay, France.

Gerke, A., Desbordes, M., and Dickson, G. (2015). Towards a sport cluster model: the ocean racing cluster in Brittany. *European Sport Management Quarterly.* doi: org/10.1080/16184742.2015.1019535.

Furthermore there is substantial research on surfing industries in Australia that should not remain unnoticed:

Stewart, B., Skinner, J., and Edwards, A. (2008). Cluster theory and competitive advantage: the Torquay surfing experience. *International Journal of Sport Management and Marketing, 3*(3), 201–220.

Warren, A., and Gibson, C. (2013). Making things in a high-dollar Australia: the case of the surfboard industry. *Journal of Australian Political Economy, 71*, 27–50.

Finally I would like to mention a very useful contemporary report on the global surf industry:

Jarratt, P. (2010). *Salts and Suits: How a Bunch of Surf Bums Created a Multi-billion Dollar Industry … and Almost Lost It.* Praharan, Vic.: Hardie Grant Books.

References

Asheim, B. T. (2000). Industrial districts: the contributions of Marshall and beyond. In G. L. Clark, M. P. Feldman, and M. S. Gertler (eds), *The Oxford Handbook of Economic Geography* (pp. 413–431). Oxford: Oxford University Press.

Babiak, K., and Wolfe, R. (2009). Determinants of corporate social responsibility in professional sport: internal and external factors. *Journal of Sport Management, 23*(6), 717–742.

Becattini, G. (2002). Industrial sectors and industrial districts: tools for industrial analysis. *European Planning Studies, 10*(4), 483–493.

Bellandi, M. (2002). Italian industrial districts: an industrial economics interpretation. *European Planning Studies, 10*(4), 425–437.

Carroll, A. B. (1979). A three-dimensional conceptual model of corporate performance. *Academy of Management Review, 4*(4), 497–505. doi: 10.5465/amr.1979.4498296.

Carroll, A. B. (1999). Corporate social responsibility: evolution of a definitional construct. *Business & Society, 38*(3), 268–295. doi: 10.1177/000765039903800303.

Chetty, S. (2004). On the crest of a wave: the New Zealand boat-building cluster. *Entrepreneurship and Small Business, 1*(3/4), 313–329.

Dolles, H., and Soderman, S. (2008). Mega-sporting events in Asia – impacts on society, business and management: an introduction. *Asian Business & Management, 7*(2), 147–162.

Dunning, J. H. (1980). Toward an eclectic theory of international production: some empirical tests. *Journal of International Business Studies, 11*(1), 9–31.

Eisenhardt, K. M. (1989). Building theories from case study research. *The Academy of Management Review, 14*(4), 532–550.

EuroSIMA. (2009). *Le marché des sports de glisse dans le monde*. EuroSIMA.

EuroSIMA. (2015a). *EuroSIMA Activity Report 2014*. Annual Activity Reports. Retrieved 12 May 2015, from http://issuu.com/eurosima/docs/rapport-2014-eng.

EuroSIMA. (2015b). History. Retrieved 30 June 2015, from www.eurosima.com/presentation-1.html.

EuroSIMA. (2015c). Presentation. Regular members. Retrieved 30 June 2015, from www.eurosima.com/presentation/EuroSIMA-Members-11-172-1.html.

Falchi, A. (2006). *Pôle glisse. Dossier de candidature du pôle de compétitivité glisse en Aquitaine*. Aquitaine, France: CRT Estia. Innovation.

Gerke, A. (2014). *The Relationship between Interorganisational Behaviour and Innovation within Sport Clusters*. Unpublished doctoral dissertation, Paris-Sud University, Orsay, France.

Gerke, A., Desbordes, M., and Dickson, G. (2015). Towards a sport cluster model: the ocean racing cluster in Brittany. *European Sport Management Quarterly*, 5(3), 343–363. doi:org/10.1080/16184742.2015.1019535.

Glass, M. R., and Hayward, D. J. (2001). Innovation and interdependencies in the New Zealand custom boat-building industry. *International Journal of Urban and Regional Research, 25*(3), 571–592. doi: 10.1111/1468-2427.00330.

Jarratt, P. (2010). *Salts and Suits: How a Bunch of Surf Bums Created a Multi-billion Dollar Industry . . . and Almost Lost It*. Praharan, Vic.: Hardie Grant Books.

Kellett, P., and Russell, R. (2009). A comparison between mainstream and action sport industries in Australia: a case study of the skateboarding cluster. *Sport Management Review, 12*(2), 66–78. doi: 10.1016/j.smr.2008.12.003.

Marshall, A. (1920). *Industry and Trade* (3rd edn). London: Macmillan (originally published 1919).

Miles, M. B., Huberman, A. M., and Saldaña, J. (2014). *Qualitative Data Analysis: A Methods Sourcebook* (3rd edn). London: Sage.

Outdoor Sport Valley. (2014). Outdoor Sports Valley et EuroSIMA unissent leurs efforts et fusionnent leurs services au profit de leurs membres [Press release].

Parker, G., and Beedell, J. (2010). Land-based economic clusters and their sustainability: the case of the horseracing industry. *Local Economy, 25*(3), 220–233. doi: 10.1080/02690941003784275.

Porter, M. E. (1998). Clusters and the new economics of competition. *Harvard Business Review*, November–December, 77–90.

Preuss, H. (2005). The economic impact of visitors at major multi-sport events. *European Sport Management Quarterly, 5*(3), 281–301. doi: 10.1080/16184740500190710.

Région Aquitaine. (2015). Invest in Aquitaine. Retrieved 30 June 2015, from www.invest-in-southwestfrance.com/board-sports.html.

Rugman, A. M. (2010). Reconciling internalization theory and the eclectic paradigm. *Multinational Business Review, 18*(2). doi: 10.1108/1525383X201000007.

Sarvan, F., Başer, G. G., Köksal, C. D., Durmuş, E., Dirlik, O., Atalay, M., and Almaz, F. (2012). Network based determinants of innovation performance in yacht building clusters: findings of the SOBAG project. *Procedia – Social and Behavioral Sciences, 58*, 830–841. doi: 10.1016/j.sbspro.2012.09.1061.

Scarfe, B. E., Healy, T. R., Rennie, H. G., and Mead, S. T. (2009). Sustainable management of surfing breaks: case studies and recommendations. *Journal of Coastal Research, 25*(3), 684–703.

Shilbury, D. (2000). Considering future sport delivery systems. *Sport Management Review, 3*(2), 199–221.

Stewart, B., Skinner, J., and Edwards, A. (2008). Cluster theory and competitive advantage: the Torquay surfing experience. *International Journal of Sport Management and Marketing, 3*(3), 201–220.

Warren, A., and Gibson, C. (2013). Making things in a high-dollar Australia: the case of the surfboard industry. *Journal of Australian Political Economy, 71*, 27–50.

Yin, R. K. (2009). *Case Study Research: Design and Methods* (4th ed.). Los Angeles, CA: Sage.

5 Surfboard making and environmental sustainability

New materials and regulations, subcultural norms and economic constraints

Chris Gibson and Andrew Warren

Introduction

Surfers are well aware of oceanic sustainability issues such as water quality and pollution, impacts of tourism, and local conflicts over coastal development. But there are also sustainability problems associated with the very equipment needed to participate in a surfing life. Surfboards are manufactured items that entail a host of upstream labour and environmental issues. This chapter accordingly discusses environmental sustainability issues in the surfboard-making industry, and dilemmas that arise as a consequence of uneven regulation, and the industry's combination of structural economic features and subcultural origins. We draw on qualitative, longitudinal research where we have visited and interviewed people in 36 surfboard-making workshops in Australia, Hawai'i and California over half a decade (see Warren and Gibson 2014). In this chapter we document sustainability issues such as dependence on petroleum products and harmful chemicals, differences in environmental regulation and poor waste management practices – issues related to making surfboards with which many surfers may not be so familiar.

Such issues are linked to the production processes involved in surfboard-making, which we describe in the first section of the chapter. The qualities of the finished product, especially its disposability and (lack of) durability, also influence overall environmental impact. In earlier eras of wood construction, exemplified in pre-colonial Hawai'i – where board-making was governed by customary practices and part of a revered craft in fine timberwork – surfboards were expected to last. That sentiment underpins the current revival in timber board-making among collectors and connoisseurs. Nowadays, though, most regular recreational surfers can go through two or three polyurethane (PU) foam surfboards every year.

In this regard, surfboards appear to be increasingly like most other mass-consumerist commodities – throwaway items with very limited functional life. Since the 1980s, as surfing became big business, surfboard-making companies with local origins have turned into corporate entities. Surfboards are the figurative heart of a wider, global surf-manufacture industry with immense power to fuel consumerism. Tentacles have spread into related retail industries and manufacture of wetsuits, apparel, shoes, sunglasses, watches and hats. Each of these consumer items – which are in turn caught up in high-throughput fashion cycles of manufacture–retail–purchase–use–dispose–replace – entails its own set of upstream sustainability issues in different countries that are rarely transparent to the consumer. As a preface to this chapter, it is therefore important at one level to consider surfboards within a wider network of surf apparel and equipment manufacture with a complex host of environmental and labour issues, spread across many countries.

Sustainability issues for surfboard-making are nevertheless also refracted by a combination of factors that pertain to the local contexts of production. Characteristics of how the industry emerged in specific places and times has led to environmental, labour and health issues within individual workshops. Later in the chapter, we describe how sustainability issues are exacerbated by the industry's highly informal, subcultural 'scenes', from which surfboard manufacturing emerged in an incremental, haphazard fashion with minimal regard for environmental impact and regulation. That regulation in turn varies across countries and states – so that well-meaning environmental protection instituted in one location can have the adverse effect of shifting the problem elsewhere.

Despite such problems, there are exemplary cases where innovators have sought to 'do the right thing', experimenting with new materials. We describe some of these advances. Elsewhere, surfboard workshops have installed new production and waste management technologies, only to be caught out by resulting problems of necessary increased production in order to cover high capital costs. The industry's strong local ties and use of hand-made production techniques provide a rich cultural heritage to the industry, but also effectively cap workshop size and capacity. Capital investments necessary to produce surfboards with smaller environmental footprints put such small workshops at risk of not being able to make or sell enough boards to survive. The result is a paradox between, on the one hand, surfboard making's 'soulful' side and increasing recognition of environmental impact, and on the other, a small-scale cottage production model that eschews regulation, and continues to use high-impact petroleum-based products that are harmful to the very environment that surfing culture cherishes.

This paradox erupted on 'Blank Monday' – 5 December 2005 – when the world's largest supplier of raw materials for the surfboard industry, Clark Foam (with a turnover in excess of US$25 million p.a.), ceased making pre-fabricated 'blanks' (from which surfboards are shaped) and began destroying long cherished moulds and irreplaceable equipment (Finnegan 2006). In a fax sent to his customers, company founder Gordon 'Grubby' Clark explained his reasons for closing:

> Effective immediately Clark Foam is ceasing production and sales of surfboard blanks.... The short version of my explanation is that the state of

California and especially Orange County where Clark Foam is located have made it very clear they no longer want manufacturers like Clark Foam in their area. The way the government goes after places like Clark Foam is by an accumulation of laws, regulations, and subjective decisions they are allowed to use to express their intent. Essentially they remove your security, increase your risk or liability, and increase your costs.

(Quoted in Warren and Gibson 2014: 107)

A specific contention was with the environmental and workplace safety consequences of making foam blanks using a variety of harmful chemicals. Tightening Californian environmental and safety restrictions had been placed on use of a toxic chemical used in polyurethane production called Toluene Di Isocyanine (TDI) in the blank casting process. Knowledge of the toxicity and environmental impact of such chemicals had lurked in the industry for decades, but little had ever been done to improve materials and production processes. Blank Monday brought the issue to the surface, and revealed to the surfing community the ugly kinds of environmental and economic issues upstream in the surfboard industry.

We discuss a range of such issues here, and survey attempts after Blank Monday to improve environmental sustainability performance in the industry. Much experimentation has transpired, but many unsustainable practices continue, amidst economic constraints and 'traditional' ways of doing things within surfing subculture.

Before we proceed, it is worth clarifying the approach taken here to our interpretation of sustainability issues: we primarily focus on *environmental* aspects of sustainability – conceptualized in the accepted manner as wise use of earthly resources while minimizing long-term ecological detriment, such that future generations can continue to enjoy those same resources. Such a focus foregrounds issues that impact upon ecological quality and human and environmental health. But we also recognize that sustainability is more accurately a complex mix of environmental pressures entwined with economic, social, political and cultural dimensions. Together these dimensions are important in explaining the environmental consequences of surfboard-making, and the possibilities (and limits) to improving sustainability performance.

For this reason, interwoven in the discussion below regarding environmental impacts are discussions of related factors such as the structure of the surfboard industry, regulation, changing technology and production methods, and the inheritance of subcultural attitudes and arrangements between makers and customers. What transpires from this interweaving of factors is that there are many paradoxes and trade-offs in surfboard-making that make it difficult to prescribe simple pathways forward. Gains made in some areas are offset by losses in others; steps that might solve one problem create new ones (cf. Head *et al.* 2013). Nevertheless we hope that by drawing together discussion of these issues and paradoxes, readers – and especially surfers themselves, who through their purchasing decisions encourage positive change in the industry – will be a degree more familiar with the key issues.

How – and where – are surfboards made?

There are two simple labour specializations in the contemporary surfboard production system, reflecting the industry's origins as a do-it-yourself backyard industry, and before that, the basic division of labour in traditional Hawaiian method. These two specializations are *shaping* and *sealing*. The *shaper* is responsible for designing and sculpting out the surfboard's profile or 'shape'. Whereas once cuts of timber were the dominant material worked upon, surfboards are now mostly made from PU foam, adapting generic blocks of material called blanks. Following almost identical methods pioneered by Hobie Alter and Grubby Clark in the late 1950s, liquefied PU is poured into concrete casts where it cures and forms a solid mass that is the blank. Moulds are set in variety of lengths and widths.

The other main materials used instead of PU foam in blank construction include expanded polystyrene (EPS) and extruded polystyrene (XPS). EPS and XPS boards last much longer than PU boards – an immediate advantage on the sustainability front. XPS is in turn much more dense than EPS, giving a tighter foam beading and a much better strength to weight ratio, but is more difficult to work, and to fix, when dinged.

Shaping workshops then order PU or EPS/XPS blanks to suit their needs from supply companies. In the United States Clark Foam was by far the dominant supplier until Blank Monday. In Australia, companies such as Burford Reinforced Plastics, South Coast Foam and Bennett Surfboards (manufacturing blanks as Dion Chemicals) are the dominant players. In southwest England, where the sport has expanded rapidly in recent years (supporting a burgeoning shaper scene) blanks are imported from the United States and Asia (see below).

Blanks are then either hand-shaped, or shaped by automated machines. Where blanks are hand-shaped, after selecting an appropriate mould the shaper traces the outline of the surfboard onto the blank (the 'plan shape'). Next a handsaw or electric jigsaw is used to cut out the plan shape from the blank. After this the shaper begins planing rougher sections of foam, working to achieve a smooth and even finish along the rails, while reducing thickness through the blank to suit the design they have created. After planing the surfboard's length, thickness and width to the desired dimensions, the shaper uses surface form tools (surform) to fine-tune each design. Features such as tail concaves and nose shapes are delicately crafted with the surform.

Where blanks are automatically shaped, computer algorithms based on hundreds of precise measurements from existing physical boards drive machines that cut shapes with a very high degree of accuracy. Frequently these mimic or even directly copy ideal prototype designs or 'magic' shapes that were previously crafted by hand shapers on existing boards. Next, with both hand-shaped and machine-shaped boards, sandpaper of different grit size helps further refine the design. In larger workshops finer sanding work is often devolved to a specialized sander, employed to ensure efficient production when factories need to move through large numbers of orders. In smaller workshops a single shaper does all the sanding work.

After the surfboard's shape is finished it moves to glassing. The *glasser* (also called a *laminator*) seals the surfboard to ensure the foam shape is waterproof and rigid. Glassers layer surfboards in fibreglass cloth, spreading liquefied resin over the top and bottom surfaces of the board to give a smooth and shiny finish. While shapers regularly receive most of the fame and attention for their work as designers and artists in the production process, glassers play an essential role in surfboard manufacturing. Mistakes here ruin the board.

The glasser's job begins with layering – called 'lapping' – the finished shape with lengths of fibreglass cloth. Next the glasser spreads a liquefied resin to begin the process of sealing the board. Workshops use two types of resin – polyester and epoxy. Epoxy resins are stronger and more adhesive to the fibreglass sheeting compared with polyester resins. But epoxy is only suited for use with EPS/XPS blanks because the resin adversely reacts with traditional PU foam, causing discoloration. Epoxy resin is also more difficult to spread over the blank and more expensive than traditional polyester resins – on average 2.5 times the price per pound. As a result EPS/XPS blanks and epoxy resins are used much less frequently than PU foam and polyester resin.

After lapping the glasser completes the fill coat, also known as a 'hot' coat. Here the resin is not actually heated but used to saturate the fibreglass cloth and fill gaps in the weave. The process is carried out on both sides of the board. Once the fill coat cures the board is again intensively sanded with different grit sizes, which helps smooth out rough bumps and imperfections. The board is then cleaned with an acetone and polished to achieve a dull finish. Like shaping, the sanding and polishing work after the resin has cured is often devolved to an apprentice glasser, specialized sander or polisher. Once the glassing is finished the desired result is an evenly covered and sealed surfboard, which is waterproof and able to withstand significant beatings from breaking waves and surfing bodies.

The typical surfboard workshop is a collection of separate spaces, divided and organized to allow the completion of different work tasks: shaping, glassing, drying, sanding and art designs are all usually completed in their own separate rooms. Glassing rooms must be well ventilated, with good lighting. In California and Australia, workshops are required to store materials (resins, hardeners, paints, solvents and acetone) in a secured room, in accordance with local environmental and workplace safety regulations.

This 'traditional' set of arrangements (at least, since the 1950s) is also rapidly changing, especially for the shaping stage. Since the 1990s, modern computerized production methods now operate within the surfboard industry using different economies of scale to traditional manual approaches. Not only have computerized shaping and design replication been used to up-scale production to meet demand from larger numbers of novice surfers, but surfboard-making companies of different sizes have increasingly moved to manufacture boards in non-surfing regions where there are cheaper factors of production, and where environmental regulations are less stringent. The focus of these new spaces of surfboard production has been Asia, especially factories in China and Thailand,

and smaller surfboard workshops in 'cheap' surf travel destinations including the Philippines and Indonesia. In such places computerized shaping technologies are being used to replicate standard designs en masse. In China, for example thousands of boards are exported weekly, most from Hong Kong, the Guangdong and Zhejiang provinces. The dominant market destinations for these boards are the United States, Australia, Brazil and Western Europe.

Diverse companies now manufacture (or subcontract the making of) surfboards, encompassing the traditional local workshops, surf corporations with countercultural origins such as Rip Curl, specialist factory producers newly present in the market (such as Global Surf Industries, Firewire and SurfTech, all based in Thailand), and diversified manufacturers such as BenPat International and SHY Technology (both based in China) who make a range of other goods beyond surfboards including golf clubs and skateboards. Where surfboard-making has been offshored, CAD/CNC technologies are used to shape boards, with thousands of models manufactured from the same design. Glassing is carried out internally within the same factory, using a Fordist production line approach.

Larger, well-known independent surfboard labels seeking to expand their export market increasingly use foreign companies as contractors to produce their boards by computer. Global Surf Industries (producing 50,000 boards annually) and BenPat International (making an estimated 30,000 boards annually) shape, label and glass surfboards for other workshops before organizing shipping to final retailers. The lower labour and overhead costs of these factories allows them to charge comparatively less for their products while maintaining higher profit margins. In these ways surfboard manufacture is becoming more like other mass-consumer industries: high throughput, automated production, disposability. Costco for instance began stocking Chinese made surfboards in California and Hawai'i from 2008, selling them for US$200 to US$300. In most cases such prices are well below the basic production costs for competing local board-makers, and the boards themselves are low quality, lasting for a shorter period of time and having to be replaced more frequently.

Despite the shift to CNC automation, hand-shaping persists in many places, and the structure of the industry – geared around local shapers and small work-shops – survives. Within striking distance of key beaches and prized breaks, groups of local surfers have sustained commercial demand for hand-made surf-boards, even with cheaper imports available. The coastal and regional distinc-tiveness of surfing's underlying geography is an insurance policy of sorts. Local knowledge of breaks and wave types matters, as does attention to the individual weight, preferences and needs of surfers. Place seeps into the design and reputa-tion of surfboards. Accordingly, environmental sustainability issues associated with surfboard-making operate concurrently at different levels associated with a diverging mix of manufacturing techniques, regulations and locations – both global and local.

Environmental issues and regulation

A host of environmental problems is associated with the above processes of making surfboards: use of non-renewal materials, carbon emissions, toxicity of petrochemicals, environmental pollution, waste disposal problems and health impacts from the production process itself.

Scant research has traced the contours of the carbon emissions associated with surfboard production. According to the most prominent study, carried out at the University of California Berkeley, the average surfboard creates 375 pounds of CO_2 emissions in the production process (Schultz 2009). Over its entire life-cycle, the carbon impact of a typical US made, 5.5 lbs shortboard is 600 lbs of emitted CO_2. Regular recreational surfers can go through two or three surfboards a year – further multiplying the overall carbon impact of making, buying and using surfboards.

The material used in most surfboards – polyurethane – is a petroleum product, and it accounts for approximately a quarter of the carbon footprint of the finished product (Schultz 2009). Blanks are mostly cast from PU foams, with resins, catalyst and acetone used for sealing, cleaning and polishing. TDI has been used in surfboard blank manufacturing since the 1960s, but is now recognized as a serious lung irritant linked to chronic asthma. Resins meanwhile account for approximately 22 per cent of the carbon impact of a finished PU board, and 37 per cent of an EPS/epoxy board. Polyester resins are also infamous in the industry for their emission of volatile organic compounds (VOCs) during the production process. Fibreglass – impregnated with resin to create the hard shell of the surfboard – accounts for approximately 5 per cent of the carbon emissions impact of surfboard manufacture. Fins meanwhile, are predominantly made from petroleum-based plastic, and entail further carbon emissions in production. Beyond carbon emissions and dependence on petroleum products, waste management is an issue in shaping workshops: one estimate is that approximately a third of all raw materials entering a typical workshop end up on the workshop floor or disposed as garbage (Staiger and Tucker 2008).

Depending on how close workshops are located to blank suppliers, there are also emissions associated with transporting blanks – although these are nowhere near as significant as the emissions that result from physical production of the blanks in the first place. When Clark Foam (which at the time supplied about 80 per cent of the surfboard market in the United States) closed on Blank Monday, the sudden downturn in the supply of blanks exacerbated short-term emissions impacts as American workshops sought blanks supplies further afield, including east coast Australia.

Although no science has been conducted that quantifies such impacts exactly, the case does anecdotally illustrate how environmental sustainability issues are entangled in complex geographies of regulation and supply: hence enforcement in one location (California) aimed at improving environmental performance can merely offset those impacts to other locations (Australia, Asia), and perversely generate other kinds of impacts such as transport-related carbon emissions needed to ensure consistency of supply. From a sustainability perspective, there are constant trade-offs such as these that are a function of the shifting geography of manufacturing goods. In time, new PU moulding factories, including US Blanks, Foam E-Z, Arctic Foam and Just Foam emerged to fill the void left when Clark and Walker Foam closed. Surfboard workshops in southern California now spread their blank orders across several suppliers to ensure there is not a repeat of Blank Monday. Nevertheless significant differences exist across jurisdictions within and beyond the United States in the degree and enforcement of environmental protection regulation.

Environmental health impacts

Our research also documented physical health problems common among surfboard-makers, from aches and pains associated with manual work, to more serious health predicaments linked to the environmental toxicity of input materials. Acute problems were especially common among older participants who have worked in the surfboard industry for long periods of time. Pre-1960s surfboard-making relied on the use of hardwood timbers, lacquers and plant-based waxes to waterproof surfboards. Although there were no such things as occupational health and safety standards back then, materials used were mostly organic and reasonably safe. In the contemporary surfboard industry commercial workshops use volatile synthetic materials and chemical components that harm workers.

Blanks contain materials with active ingredients or components that are irritating to the body and in some cases harmful to long-term health. The foam used in surfboard blanks is composed of fine reactive polymer compounds. When shaping by hand or using automated machines the blank releases small particles of polyurethane or polystyrene foam into the surrounding air. If inhaled, foam particles can become blocked in airways and cause respiratory illness or inflammation of the airway.

Likewise, glassers use liquid resins to fill and finish coat boards, catalyst to harden the resin and acetone to polish and clean up spills. Dangerous fumes are easily inhaled and glassers regularly come into physical contact with potentially harmful chemicals. Formaldehyde helps the resin absorb deeply into the fibreglass cloth. Over the last decade several medical studies have examined the exposure of workers to formaldehyde across a number of different industries (including funeral workers who used formaldehyde to embalm bodies), and there are alarming health problems resulting from regular and prolonged exposure (Hauptmann et al. 2009). Among workers using formaldehyde is an increased frequency in cases of, and mortality from, myeloid forms of leukaemia. According to one medical study prolonged exposure to formaldehyde heightens the risk of contracting cancers of the hematopoietic and lymphatic systems – particularly myeloid leukaemia, which affects the bone marrow (Beane Freeman et al. 2009). Surfboard-makers work with or are in close proximity to dangerous chemicals and hazardous materials: the resins, catalyst, glues, paints, acetones and inevitable clouds of foam dust that permeate every factory. Such substances are often inhaled in small quantities every day or come into direct contact with the body. Asthma and respiratory complaints were the most frequently discussed among our interviewees, considered the result of extended exposure to fumes released by resins, catalysts and acetones.

New materials and possibilities

At the time of writing the surfboard industry continues to use predominantly PU foam and fibreglass for surfboard-making – accounting for up to 90 per cent of commercial production. The Blank Monday episode, followed in 2007 by the liquidation of another large blank supplier, Walker Foam, sparked surfboard-makers to experiment with the use of different types of foam and resin combinations, recycled ingredients and bamboo and hemp cloths. Beyond concerns over speed and strength, some new construction materials are also being sought for environmental reasons. In some workshops, trialling different materials is being driven by an environmental and health consciousness relating to PU foam and resin in combination with attempts at reducing weight and creating more enjoyable boards.

Much of the innovation in ecologically sustainable surfboards has emerged from California – where environmental regulation has been strictest. Following the Blank Monday crisis, two brothers Rey and Desi Banatao (who both had materials science degrees) founded Entropy Boards in Santa Monica, California. Together they developed a new recipe for a 'bio-board', in which sugarbeet oil replaced polyurethane, and hemp cloth replaced some of the fibreglass in the glassed shell (Stone 2008). Sugarbeet oil is almost identical to PU in a chemical sense, but in processing uses less toxic chemicals. Under the brand name Super Sap, Entropy now markets epoxy resins partially made from the waste by-products of the pulp, paper and biofuels industries, with biological content between 25 and 50 per cent of the material. It claims reductions in overall carbon footprint of at least 50 per cent compared with traditional petroleum resins. Other workshops have meanwhile experimented with recycled polystyrene and alternative ingredients, and fins made from bamboo and recycled plastic and carpet that are glassed-in permanently to the board (rather than made in such a way that they connect to the board using plastic fin boxes). PU foam sourced from algae has been developed and adopted by Arctic Foam (sponsors of John John Florence), who have plans to produce on a commercial scale this year.

Supporting such experimentation, a new labelling scheme, ECOBOARD, has been introduced by a new non-profit benchmarking agency, Sustainable Surf (established in 2011 by Michael Stewart and Kevin Whilden). Akin to food and energy star labelling schemes, ECOBOARD aims to provide more transparent information to consumers that purchase certified boards that have been produced in a manner that minimizes impacts on workers and the environment. Verified ECOBOARDS must be made from blanks containing a minimum of 25 per cent recycled foam or biological content; with resin made from a minimum of 15 per cent bio-carbon content with low or zero VOCs; or made from an alternative material altogether, principally wood, which reduces or eliminates dependence on petroleum-based foam and resins (www.sustainablesurf.org/ecoboard/benchmark/). At the time of writing ECOBOARD had been endorsed by the US-based Surf Industry Manufacturers Association (SIMA), and some 33 workshops had signed up to produce boards using the ECOBOARD labelling and accreditation (Bradstreet 2013). Of these, most were in southern California, clustered in the heart of the industry's established territory: in San Clemente, Oceanside and San Diego. A scattering of other workshops make ECOBOARD certified surfboards on the US east coast, as do a couple in southwest UK. To the best of our knowledge, none are registered in Australia.

Also from the southern California hub, the 'Waste to Waves' programme encourages consumers to collect and recycle EPS foam (typically found as packing material when purchasing a new TV or appliance) via surf shops who host collection boxes (see http://wastetowaves.org/2011/11/how-it-works/). EPS materials are then collected by a company called Marko Foam – a blanks manufacturer who retrieves the material when delivering new surfboards to the same shops – and reprocessed into surfboard blanks. This process still requires energy to blow the EPS material into a blank (hence generating carbon emissions) but, it is claimed, significantly reduces the roughly 70 per cent of impacts in the production of virgin EPS that come from the extraction and processing of raw materials used to make foam (www.sustainablesurf.org/ecoboard/technology/).

In the southwest UK, a conglomerate of composite materials manufacturers have formulated resins for a PU foam core blank ('Ecoblank') made from 40 per cent castor oil, as well as a UV-cured resin system ('EcoComp UV-L resin') with more than 90 per cent linseed oil content (Staiger and Tucker 2008). Resulting ECOBOARDS, made locally, use hemp fibre instead of fibreglass. The manufacturers claim that the ECOBOARD contains 55 per cent renewable content, and that the linseed UV cure system wastes less resin than conventional polyester due to the long pot life of the substance (the curing process does not begin until the resin is exposed to UV light). Using these materials, workers are not exposed to VOCs and acetone is not required for clean up.

Meanwhile, the revival of wooden surfboards – fuelled by the growing 'retro' movement and vintage surfboard collector scene – has provided another alternative. Full lifecycle analysis suggests that timber production results in fewer CO_2 and other environmentally damaging emissions than PU foam boards (Hole 2011). Timber construction also reduces petrochemical dependence and eradicates VOC toxicity issues involved in PU (Grees 2014), but it does introduce new irritants associated with wood dust. Wood dust is much less toxic, but still requires everyday management. Glues and lacquers are still used.

Moreover, wood entails a different set of upstream issues, connecting surfboard-making to forestry management and transport-related issues associated with the timber trade. Among the timbers commonly used in contemporary surfboard-making are balsa (*Ochroma pyramidale*, sourced overwhelmingly from Ecuador), which featured heavily in earlier eras of board manufacture.

A fast-growing, short-lived tropical tree, supplies of naturally grown balsa became strained in the 1950s and 1960s with the boom in surfing, alongside other parallel industrial uses (such as in aircraft manufacture). Nowadays balsa is plantation grown in Ecuador, harvested after six to ten years of growth. It is not listed as threatened ecologically according to the Convention on International Trade in Endangered Species (CITES) or the IUCN Red List of Threatened Species. It is, however, expensive – limiting commercial applications.

An alternative, paulownia (*Paulownia tomentosa*), is increasingly being used in surfboard-making. It too is fast-growing and light, and not threatened according to CITES or the IUCN Red List – and can be harvested in some cases after only five years. In some areas it is even considered an invasive species. Surfboard-makers in Australia are increasingly turning to paulownia, and some interviewed in this research had recently become involved in plantation cultivation of paulownia in Far North Queensland (which has suitable climate, rainfall and soil), in order to improve supplies. Tom Wegener, a well-known Australian shaper renowned for advances in wooden surfboard construction, uses paulownia (he is also at the time of writing completing a PhD on sustainability issues associated with surfboard-making). The market for wooden boards is growing, though is unlikely to replace composite fibre surfboards given manoeuvrability and performance problems. It nevertheless is becoming a viable niche, and one that connects surfers to both the culture's traditional heritage in Hawai'i, and a more 'organic' relationship with the environment.

Subcultural origins and economic constraints

Despite growing environmental and health awareness, and experimentation with new and alternative materials, many unsustainable practices and dangerous behaviours within workshops continue, and magnify risks. In this final discussion in the chapter, we argue that inhibiting sustainability improvements are factors linked to the industry's informal DIY origins, which has given rise to a distinctive – and limiting – mix of economic structure and subcultural norms.

Pioneer surfboard-makers began in the heady days of the 1940s and 1950s. Surfboard-making scenes were do-it-yourself pseudo-industries operating first from Hawaiian beaches, and then out of garages and sheds in California and Australia. Informal, experimental and almost completely unregulated, board-making became a part-time accompaniment to days spent surfing, drinking and hanging out. Surfboards were made out of necessity, rarely with business acumen. Early surfboard-making was characterized by coastal cultural life and small-scale 'backyard' production. After the Second World War, surfing – particularly its Californian variant – began its progression towards mainstream social acceptance and Western consumerism. As more people took up surfing in the 1950s and 1960s, and as tourism in all three Pacific regions boomed, the market for surfboards grew locally. Several early surfboard-makers found they could make respectable livings from crafting boards for local waves. Early innovators became renowned 'legends' of the sport and master craftspeople.

Subsequent generations of innovators and hand-shapers followed and reflected generational changes in preferred surf breaks and styles.

The resulting economic structure is one where a small number of 'lead' firms fuel technical (and in the context of this chapter, environmental) innovation in board design, while a much larger number of comparatively anonymous, usually smaller-scale operators and local do-it-yourself board-makers satisfy demand for surfboards tailored to local conditions. Their boards might never feature in pro-tour competitions yet they are very much a part of surfboard-making as a grass-roots industry.

In regard to environmental sustainability, this structure is reflected in the degree of experimentation, adoption of alternative materials and persistence of old 'habits'. In southern California, where environmental consciousness, stricter regulation and sheer market size are combined, there are a number of lead firms pioneering sustainability initiatives. Elsewhere, and especially in Australia, sustainability has lagged behind, and PU foam boards remain stubbornly the norm.

Complicating this picture is that local distinctiveness and close connection to a subcultural scene limits both the size of surfboard-making firms and their capacities to invest in new materials, techniques and facilities. Prioritizing relationships with local customers, shaping boards by hand and customizing orders are all much-cherished characteristics of the traditional format, but they do limit business expansion, and thus the capacity to leverage debt in order to invest in new, more sustainable materials and facilities.

One Australian workshop interviewed for this research (for whom we wish to protect anonymity) put in place 'best-practice' environmental and workplace safety and conditions in the early 2000s, by investing many hundreds of thousands of dollars in a purpose-built, ground-up designed production facility. Capacity expanded accordingly, but debt taken on board in order to finance the best-practice operation exposed the firm to greater levels of risk. Higher volumes were necessary to meet debt repayments, which meant increasing reliance on marketing and advertising, engaging in dodgy retail consignment agreements and 'ghost shaping arrangements' (whereby they shaped unbranded boards that were subsequently sold in megastores with big brand decals), and higher risk strategies to expand export markets beyond the loyal local surfing community.

This workshop's trade went well initially, but then soured with the global economic downturn and a (then) high Australian dollar. The firm could not keep up with debt repayments, and sold their state-of-the-art facilities at a considerable loss (upwards of US$200,000). They subsequently downsized the business – moving to a more modest space rented alongside another surfboard-maker, without the high-end best-practice environmental features, and focusing their product only towards the local surfing crowd. This failed experiment in up-scaling production with best-practice facilities demonstrates the risks associated with stretching beyond a traditional local, craft base. In surfboard-making, on-going viability depends on tight social relationships between makers and customers, even if that means production (and profit) remains ultimately constrained. The very same constraints prevent substantive investment in new plant, machinery and alternative materials needed to improve environmental sustainability performance.

Meanwhile, despite many advances in new and alternative materials, the dependence on PU foam and polyester resins remains remarkably difficult to shift – a consequence of both shapers' and surfers' preferences for that combination of materials. Alternative materials still struggle for legitimacy and credibility within the subculture – a direct parallel to other culturally-based industries such as guitar-making (Gibson and Warren, 2016) where a set of entrenched expectations among user groups guides the 'tradition' of manufacturing an unchanging 'type form' (Molotch 2005). In other words, when a successful 'formula' develops for how a product should look and be made, it can prove very difficult to revolutionize.

This explains why ECOBOARDS made from recycled EPS have struggled against the traditional PU, with many surfers preferring the 'feel' of PU. For elite surfers – and the millions that follow them – surfing is in essence about wave performance: 'A Nascar driver doesn't particularly care how many miles per gallon his souped-up Chevy Malibu gets, and surfers likewise obsess about speed. For half a century that has meant building boards out of polyurethane' (Woody 2012). There is even a scientific basis for the on-going preference for PU: testing on the flexural qualities of alternatives confirms poorer performance. Hemp cloth laminates fail to protect cores as well as fibreglass, and bio-foam has lower core shear strength in comparison to PU foam (Johnstone 2011). Despite growing environmental awareness among surfers, Marko Foam's Envirofoam blank makes up only 10 per cent of its total sales (Woody 2012). As Clay Peterson, an owner of Marko Foam, simply puts it, 'It's difficult to get some of those old-timers to switch and embrace the new technologies' (quoted in Woody 2012).

Old habits also die hard within surfboard-making workshops. Until 20 years ago many shapers and glassers did not wear dust or breathing masks. Pungent chemical fumes were frequently inhaled for the entire duration of the working day. This was typical in an era when surfboard-making was highly informal, unregulated and operated out of backyard sheds and garages. Frequency of exposure was high. Tony, a glasser on the east coast of Australia, was oblivious to such dangers:

> It was stupidity when I think about it now, and I get angry at myself. But at the time, you were busy glassing away and after a while you don't even smell the fumes. Resin has no odour to me anymore and I have become totally desensitized. It took someone to walk in here one day and they said to me 'put a fucking mask on' because the resin is really thick and strong. I realized I didn't even smell it. That was the problem, my sense of smell is now nearly gone.

Safe work practices in surfboard workshops often took a back seat to the time demands of finishing a new board, as Dean, a glasser in Australia, explained:

When I started you wouldn't always bother putting a face mask or a respirator on when you were glassing. Taking it off, putting it back on, wiping away the sweat because it was bloody hot. You were just thinking about getting the board done, you know. Inhaling all those chemicals; I mean even the resin we applied we found out that formaldehyde was the active ingredient. When you got it on you, you would get a burning sensation around your eyes and it made your throat sore to breathe. Where it touched your skin would be all red spots. Fuck, I mean that is a pretty good sign you're doing some damage to yourself isn't it? And here we were with it covered all of us, bloody breathing it in.

Dean's previous employer sourced their resin from a local chemical supply company that still used formaldehyde as an active ingredient. It was the cheapest option. The link between the chemical and forms of cancer have started to worry him:

I can't help but think about it [getting sick]. It worries me a lot actually. I feel like the clock is ticking; you've got to try and put it out of your mind but I've read things on the Internet that explained the chances of getting cancer and that does really play on my mind. It's just the reality of it, being so naive to the dangers of what you're doing.

In the United States significant steps have been made in California by the State Fire Department and Environmental Protection Agency (EPA) to restrict the use of phenol and formaldehyde as ingredients in the manufacturing of resin, along with TDI. Yet elsewhere restrictions are less stringently enforced.

Meanwhile the surfboard industry still lacks consistent and clear occupational safety guidelines. While safe work inspectors and local environmental protection agencies in most major surfboard-making regions now carry out regular checks, safety standards we observed in workshops still vary considerably. At one factory we visited in Australia large drums of resin and two tubs of acetone sat near the feet of glassers, waiting to be knocked over. Ironically the owner explained how the state regulatory authority overseeing worker safety had recently ordered the company to better ventilate the glassing room and construct a quarantine space where drums of flammable chemicals could be stored. The workshop was threatened with a large fine if they did not implement the changes before the next inspection. However, the regulatory officer was a friend of the owner and had informed him of the next inspection date, meaning the job was not an immediate priority.

In another example, a young employee on O'ahu who was responsible for polishing and preparing boards had to abruptly leave his job under doctor's advice because the regular exposure to polyurethane foam dust and strong resin fumes in the factory badly inflamed his asthma. The workshop had been told on two separate occasions to install better ventilation in their shaping areas and glassing rooms to reduce unnecessary exposure to foam dust. While 'old' work routines are slowly changing as shapers and glassers become more aware, significant damage has already been done, and the laid-back subcultural atmosphere surrounding the typical local workshop negates attempts at strict enforcement. In Hawai'i workshops are very loosely inspected, especially those operating in home garages and backyard set-ups. Workshop owners have a relaxed attitude towards changing procedures for production. In Australia the onus is on individual workers to wear protective masks and equipment, with policing of the workshop space relying on a vigilant owner or manager. Nevertheless, precautions and perceptions may well be gradually changing. In workshops, ventilation systems are now nearly ubiquitous, and face masks more frequent in daily use.

Meanwhile, some shapers are gradually discovering that there can be a 'sweet spot' between environmentally sustainable materials and board performance. William 'Stretch' Riedel, a well-known shaper from Santa Cruz, CA, now only uses EPS blanks and bioresins. Yet, he says,

> I really didn't start doing this to be green at all. I really went for performance, as I could make lighter, stronger boards. The performance of the materials is really what makes this the greenest. It's really hard to break this board, so the customer is buying one board instead of two or three boards over time.
>
> (Quoted in Woody 2012)

In time other workshops new designs that combine structural and environmental advantages, and that are acceptable within the subculture, are slowly emerging. Michel Bourez's victory at Sunset in the 2014 Vans Triple Crown of Surfing was the first high-level victory on a board with ECOBOARD badging. Kelly Slater's purchase of a majority share of Firewire and use of EPS boards in contests (Snapper, Bells, J-Bay) combines with a developing discourse amongst commentators that EPS boards are lighter, and more responsive, suited to glassy, smaller conditions. These are early examples, but they suggest that a combination of environmental and elite sporting performance is possible, and may in the future become the norm.

Conclusions

Surfboard manufacturing has the potential to become a model industry geared towards minimizing environmental impacts. The ECOBOARD certification programme and development of new materials and recycling schemes demonstrates a willingness among some industry figures to do the right thing. Pioneers in sustainable surfboard-making have gained traction linking to issues of carbon emissions and climate change, and appealing to surfers directly in terms of related consequences such as sea level rise, ocean acidification and coral reef extinction.

Nevertheless, there are paradoxes and trade-offs. Wooden boards are the most 'natural' and least dependent on petroleum products, but perform less well in the surf. CNC shaping in larger facilities enables workers to avoid repetitive injuries and usually comes with better ventilation that reduces exposure to dust. But CNC technology deprives hand-shapers of work in an industry where valuable hand-skills were developed over decades.

While tighter environmental regulation is a positive step for ensuring the health and safety of new and future workers in the industry, it varies geographically, shifting the problem from one jurisdiction to another. Awareness is patchy, filtered by the informal subculture surrounding surfing scenes in local communities, and has come too late for others now suffering health problems due to long histories of improper work practices and unsafe factory environments.

Beyond some obvious and low-overhead measures that ought to be adhered to and enforced (ventilation, face mask-wearing), improvements in surfboard sustainability have not been easily forthcoming. Environmental sustainability is not so much prevented by lack of care for the environment or apathy in the industry, than by intersecting technical, regulatory, cultural and economic factors that shape and constrain possibilities.

References

Beane Freeman, L.E., Blair, A., Lubin, J.H., Stewart, P.A., Hayes, R.B., Hoover, R.N. and Hauptmann, M. (2009) Mortality from lymphohematopoietic malignancies among workers in formaldehyde industries: the National Cancer Institute cohort. *Journal of the National Cancer Institute* 101, 751–761.

Bradstreet, K. (2013) The ECOBOARD project founders on eco-certified surfboard materials and new SIMA partnership. *Transworld Business*, 21 February, http://business.transworld.net/122579/features/the-ecoboard-project-founders-on-eco-certified-surfboard-materials-new-sima-partnership/, accessed 16 October 2015.

Finnegan, W. (2006) Blank Monday. *The New Yorker*, 21 August, www.newyorker.com/magazine/2006/08/21/blank-Monday, accessed 16 October 2015.

Gibson, C. and Warren, A. (2016) Resource-sensitive global production networks: reconfigured geographies of timber and acoustic guitar manufacturing. *Economic Geography* 92(4), 430–454.

Grees, T.H. (2014) *A Wooden Alternative: Examining the Environmental Impact of the Production of Surfboards.* Environmental studies thesis, Bates College, http://scarab.bates.edu/envr_studies_theses/31/, accessed 16 October 2015.

Hauptmann, M., Stewart, P.A., Lubin, J.H., Beane Freeman, L.E., Hornung, R.W., Herrick, R.F., Hoover, R.N., Fraumeni Jr., J.F., Blair, A. and Hayes, R.B. (2009) Mortality from lymphohematopoietic malignancies and brain cancer among embalmers exposed to formaldehyde. *Journal of the National Cancer Institute* 101, 1696–1708.

Head, L., Farbotko, C., Gibson, C., Gill, N. and Waitt, G. (2013) Zones of friction, zones of traction: the connected household in climate change and sustainability policy. *Australasian Journal of Environmental Management* 20, 351–362.

Hole, B. (2011) *An Environmental Comparison of Foam-core and Hollow Wood Surfboards: Carbon Emissions and Other Toxic Chemicals.* Forestry thesis, University of British Columbia. https://elk.library.ubc.ca/handle/2429/36219, accessed 16 October 2015.

Johnstone, J. (2011) Flexural testing of sustainable and alternative materials for surfboard construction, in comparison to current industry standard materials. *The Plymouth*

Student Scientist 4, http://studentjournals.plymouth.ac.uk/index.php/pss/article/view Article/151, accessed 16 October 2015.

Molotch, H. (2005) *Where Stuff Comes From.* New York: Routledge.

Schultz, T.C. (2009) *The Surfboard Cradle to Grave Project.* Master's thesis, Berkeley, University of California.

Staiger, M.P. and Tucker, N. (2008) Natural-fibre composites in structural applications, in Pickering, K. (ed.) *Properties and Performance of Natural Fibre Composites.* Cambridge UK: Woodhead; Elsevier, 269–300.

Stone, A. (2008) Green wave. *Forbes,* 14 August, www.forbes.com/forbes/2008/0901/058. html, accessed 16 October 2015.

Warren, A. and Gibson, C. (2014) *Surfing Places, Surfboard Makers: Craft, Creativity and Cultural Heritage in Hawai'i, California and Australia.* Honolulu, HI: University of Hawai'i Press.

Woody, T. (2012) Surfing's toxic secret. *Forbes,* 19 April, www.forbes.com/sites/todd-woody/2012/04/19/surfings-toxic-secret/, accessed 16 October 2015.

Part IV

Informing policy domains

6 Surfing voices in coastal management

Gold Coast Surf Management Plan – a case study

Dan Ware, Neil Lazarow and Rob Hales

Introduction

Participation levels in surfing at the surfing hotspots of the world have never been higher. It is estimated that there are 35 million people in the world who consider surfing one of their recreational pastimes (*The Economist* 2012). The mainstreaming of surfing culture through increased marketing of competitive surfing events and the commodification of surf culture has led to a rise in the popularity of surfing. Coupled with population increase and decreasing costs of surfboards means that there are even more people in more places that have fewer constraints to participate. Surfers throughout the world see the result of this as overcrowding.

For Australia's Gold Coast climate and tourism marketing combine to lure surfers to live and visit by the tens of thousands. A quick glance at the homes of surfing world champions indicates that the combination of wave quality, wave frequency and competition in the line-up has bred multiple generations of world-class surfers.

In addition, to the production of high-performance surfers there are other benefits of crowding to a modern capitalist society – busy car parks, beaches and surf breaks – are all signs of opportunity for consumer spending. Vendors of surfboards, board shorts, wax, meat pies, gluten free veggie burgers, cans of coke, accommodation can all be seen as benefiting from greater numbers of surfers at surf breaks.

The economic importance is also reflected in the destination marketing for the Gold Coast where the image of surfing is used to attract tourists to the city. Additionally economic development strategies supported by state and local government have been implemented to enhance the surfing industry and one of the more visible strategies is the Queensland Government financially contributing to the annual world tour surfing on the Gold Coast.

Increasing numbers have also contributed to the power and success in campaigns surrounding surfing issues. Surfers have a long and proud history of protest and advocacy on issues that reflect the connection of surfers to the coastal and marine environment. There have been many successful campaigns against direct threats to surf breaks where surfers and surfing groups have been unified against development proposals that risk the degradation or destruction of waves (see Lazarow 2010 for a detailed description of the history of surf advocacy). As the authors will describe, the Gold Coast has been a focal point for campaigns against inappropriate coastal development for some decades now – a situation which is contributed to by the combination of the region's strong economic development imperatives against a desire by many to maintain the quality of the waves.

While there are economic benefits of increasing the number of surfers in the ocean there are also social benefits as surfing opens up opportunities for communities and classes to interact, laugh and share. There is, however, an important flipside to the surfing participation equation. The number of waves and surf breaks available to accommodate a given surfing population is limited. The implication of this is that as surfing populations increase, the quality of the surfing experience in many locales tends to decrease as a result of increased competition for waves.

Increasing participation, however, creates issues for the surfing experience – often increasing risk of personal injury, lessening the overall surfing experience and lowering the number of waves an individual might ride in any given session. Education campaigns to promote informal rule or lore systems of surfing etiquette (e.g. no dropping in) have been promoted in an attempt to mitigate conflicts between users.

Lazarow *et al.* (2007) present a typology of Surfing Capital to capture the range of issues that may affect surfing and the surfing experience. Distilled, the four categories are: wave quality, wave frequency, environmental and experiential. In 2010, this was extended to better understand the array of strategies available to manage Surfing Capital, which essentially revolve around supply and demand options. This is presented in Table 6.1.

While the more direct threats to Surfing Capital such as the degradation of supply (e.g. loss of surf break due to a marina development), less overt issues such as increasing demand present problems which traditional surfer advocacy/ protest tactics may not be able to address. There is limited evidence of the existence of a social contract, which may act to guide public policy decisions regarding (1) surfers interactions with other surfers and other users, (2) conflict with other land/sea use and (3) surfing's role in the conservation of coastal environment. As a result of the lack of a social contract, decision making regarding the use and development of coastal Australia often remains highly contested and can be considered an arm wrestle between commercial and social/ environmental interests (Moote *et al.* 1997; Rockloff and Lockie 2004; Vanclay 2012).

This chapter examines the agenda-setting processes, which led to the development of the Gold Coast City Surf Management Plan. In contrast, to many local government, management planning processes, the plan was instigated by a coalition of community surfing organisations and interested individuals as a mechanism to support the transition to a more institutionally recognised and socially acceptable social contract for surfing on the Gold Coast.

Table 6.1 Strategies to manage user impact and resource base at surf locations

Do nothing	Legislate/regulate	Modify the resource base	Educate/advocate
• Do nothing	• Restrict users through strategies such as payments, restricted access or parking, craft registration, restricted time in the water • Modify user behaviour using legislation such as requiring proficiency to surf particular areas or policing a surf break on jetskis • Community title (for example, Tavarua) • Declaration of surfing reserves	• Groynes • Seawalls • Artificial reefs • Sand bypass systems • Beach and nearshore sandbar grooming • Nourishment campaigns • Break becomes unsurfable due to water pollution	• Code of ethics (that is, road rules for the surf) • Signage • Education strategies • Surf rage, aggression, intimidation • Self-regulation/ localism • Lore • Declaration of surfing reserves • Direct action • Protests and demonstrations • Lobbying and the promotion of alternative strategies • Provision of new information

Source: Lazarow (2010).

To understand how such a contract has developed the multiple streams theory of agenda setting in public policy will be used. Kingdon's (2011) multiple streams theory of agenda setting in public policy describes three process streams: politics, problem recognition and the formation and refining of policy proposals. The theory argues that the three processes operate independently; however, on occasions they merge either coincidently or through the concerted efforts of individuals referred to as 'policy entrepreneurs'. The merging of the processes indicates the point where there are significant changes and developments in the policy agenda of a particular policy issue. This theory builds on the garbage can model of organisational choice by Cohen *et al.* (1972), where organisations are characterised as 'organised anarchies' or garbage cans where 'collections of choices looking for problems, issues and feelings looking for decisions situations in which they might be aired, solutions looking for issues to which they might be the answer and decisions makers looking for work' all circulate and occasionally align. The three streams of problems, policy and politics will now be explored and this will be followed by an analysis of these streams to highlight how the Gold Coast Surf Management Plan came into being in its present form.

The method used in this chapter is a case study approach with an embedded methodology (Dredge *et al.* 2013). All authors have been part of the policy landscape of surfing on the Gold Coast and have surfing, tourism and coastal management as significant areas of their research focus.

Problems stream

Within the multiple streams framework problems are policy issues, which attract the attention of the policy system. A policy problem, which attracts considerable attention is economic development. On the Gold Coast the local government authority, City of Gold Coast, takes an active role in stimulating economic development through its economic development unit. The recent cruise ship proposal was supported by economic development policies but is considered a problem for surfers and other interest groups.

The Spit cruise ship terminal

The 2012 Broadwater Marine Project is a response by the Gold Coast Mayor Tom Tate to the economic development 'problem' the Gold Coast faces. The following quote from Tate at the launch of the Broadwater Marine Project highlights the linkage between the economic development and the project.

> This is the opportunity to transform currently unused land into a one-of-a-kind marine-based hub in the heart of Australia's leading tourist destination. My promise to the people of the Gold Coast is to boost the economy through tourism and jobs. Together we can achieve this and create something special.
> (Tom Tate as quoted by City of Gold Coast 2012, 2 November)

The Broadwater Marine Project was the second time a Government had sought to undertake the development of a cruise ship terminal on the same area of public parkland immediately to the south of South Stradbroke island, home to a popular surf break known as TOS. The original proposal in 2004 was strongly opposed by the local surfing community on the basis that it would restrict access to TOS and that the dredging required to establish the terminal would interfere with the unique coastal processes which contribute to the high wave quality at TOS. After a lengthy campaign the state government abandoned the proposal in 2007 on the basis that it was both environmentally and economically unfeasible. The 2012 proposal was quietly announced on the eve of the 2012 Gold Coast City Council elections by successful mayoral candidate Tom Tate. The proposal by the mayor was subsequently supported by the state government who initiated a special project.

The 2012 proposal was immediately recognised as a threat by surfers, not just to the quality of waves at TOS but also to the environmental and social values of the area. Two world champion surfing figures entered into the public debate.

> Tom Tate the Mayor for the Gold Coast is proposing the construction of a cruise ship terminal that would gut The Spit causing mass impact to the marine life, waves and lifestyle on the Gold Coast.
> (2013, 2009, 2007 World Surf League (WSL) champion Mick Fanning as quoted at www.mickfanning.com 2012, 18 September)

@Goldcoastmayor Tom Tate, you're tripping, Mate (with that Spit proposal). #TooShallow and #GonnaRuinGreatSurfSpots.

(11-time WSL champion Kelly Slater as quoted at www.twitter.com
@kelleyslater 2012, 12 September)

In examining the problems stream it is important to recognise that the problems which attract the attention of policy systems may not be those same issues which are recognised as problems by a user group such as Gold Coast recreational surfers. The following quotes highlight the difference in the appreciation of the issues associated with the development of a cruise ship terminal at a surf break between the policy makers and the surfing community

He's (Mick Fanning) a good surfer but I tend to listen to people with qualifications and information to add to the debate.

(Tom Tate as quoted by Killoran 2012, 19 September)

I welcome them to come and I just want to point out the surfers there's 65 kilometres of beach here on the Gold Coast and take a pick.

(Tom Tate as quoted by Berkman 2012, 3 August)

The surfing community aren't happy about the cruise ship terminal because it will lose them some of the surf break.

(Tom Tate as quoted by Anon 2012, 16 August)

At the time of publication the proposal to construct a cruise ship terminal had lost the support of the state due to a change of government. However, the mayor had announced a revised version of the cruise ship terminal as one of his key re-election policies.

Crowding

The problem of crowding at surf breaks has traditionally gone unrecognised by the policy system. According to 11-time world champion Kelly Slater, Snapper Rocks at the southern tip of the Gold Coast is the most crowded surf break in the world.

The crowds here are like nothing I've ever seen in the world.

(Kelly Slater as quoted by Greenwood 2014, 6 February)

The ironic feature of this statement is that the promotion of professional surfing and competitions are part of what attract people to the Gold Coast with the intention of surfing, which exacerbates the crowding situation. In announcing a sponsorship of the 2013 World Surfing Championship Quicksilver Pro Snapper Rocks Event, Queensland Tourism Minister Jann Stuckey MP had the following to say:

This is the first stop on the 2013 World Championship Tour and is expected to draw big crowds over the next two weeks.... Last year's Quiksilver and Roxy Pro attracted more than 41,000 people over nine days of competition – boosting the local economy by $6.5 million.

(Stuckey 2013, 28 February)

Well respected local and former longboard world champion Wayne Deane provides the following explanation of the various factors which result in the over-crowding reported by Kelly Slater, highlighting connections between the sport, industry and government policy.

Since the inception of 'TRESBP', waves in the Coolangatta area have become more consistent because of the sand being delivered to the point. Add to that surf cams, mobile phones, the Quiksilver Pro, cheap flights, Chinese imports, Thailand imports, 50-odd board builders, and there you have a recipe for chaos.

(Wayne Deane quoted by Quinlivan 2014, 10 December)

While surf break crowding is recognised as a problem for surfers, the absence of recognition by the policy system can actually exacerbate the problem for surfers. One of the points which Wayne Dean stated above as contributing to surf break crowding are supported by the Queensland State and City of Gold Coast on the basis that they will attract crowds to the city. The problem recognised by the policy system in seeking to attract crowds is the need to stimulate economic development. From this perspective crowds are a positive outcome of government action.

Currumbin Alley

The Queensland State Government response to the death of a surfer at Currumbin Alley in 2011 provides an example of how the problem framing by the policy system can often differ from that of the local recreational surfing community with potentially significant consequences for surfers.

Currumbin Alley is a surf break that breaks across the entrance to Currumbin Creek, a popular bar crossing for recreational boat users. In May 2011 a local surfer was killed when, as he was duck diving beneath a wave, he was struck in the forehead by the propeller of a boat that was trailing the same wave into the creek. Following the death the Queensland State Government's Department of Transport conducted an investigation into navigational safety at the creek. The findings of the study were released in June 2011 and proposed a series of recommendations; of particular concern was the recommendation regarding legislation. An extract from the recommendations section is provided:

Legislation should be clarified, the current lack of uncertainty regarding whether a surfer is a vessel is problematic. Legislation should be amended, as necessary, to clearly define surf craft as either vessels or swimmers. This preliminary report offers the suggestion that the former would be preferable.

(Maritime Safety Queensland 2011: 14)

The recommendation that surfers be classified as vessels has a number of serious implications for surfing not just at Currumbin Creek but across Queensland. For instance, once classified as a vessel, surfers would have obligations to give way to vessels crossing the river bar. One possible (but hopefully extreme) interpretation of this may result in surfers being forced to exit the water each time a boat wanted to exit or enter Currumbin Creek. One has to question why Maritime Safety Queensland (MSQ) didn't consider the experience from Byron Bay, where, while the volume of vessel traffic is far smaller, the simple sounding of a horn to warn surfers of the approach of a vessel has been used to reduce the risk of collisions.

This highlights the competitive nature of problem framing, particularly where surfing interests are in conflict with other coastal uses such as boating. In this instance the problem as defined by the Department of Transport is not the safety of surfers but the capacity to regulate navigation. Fortunately this recommendation is yet to be implemented.

Policy stream

Within the multiple streams framework the policy stream is where solutions emerge. In order for the streams to merge, problems must be connected to solutions which suit the political climate. While the majority of policy proposals are advanced as a solution to a problem, the important point is that unless that problem is recognised by the policy system, the proposal is unlikely to become an agenda item. An example of this was a proposal to floodlight surf breaks to reduce crowding by extending the surfing hours into the night. The Gold Coast mayor illustrates the gap between the policy solution and the problem.

[I]t is more important to light up the junior soccer fields and other courts where children are playing. These ideas, along with the tax on surf, are just not on my radar.

(Tom Tate quoted by Anon 2014, 19 November)

Kirra Groyne

Through the 1980s and 1990s Kirra Point, already significantly but fortuitously modified through engineering works over many decades, was recognised as one of the best right-hand point breaks in the world and was the home break of a number of world champion surfers.[1] In 1996 Kirra Groyne was shortened by 30 m to improve the longshore transport of sand from Coolangatta to Kirra. With the reduction in the length of the Groyne and the commencement of sand pumping to snapper rocks by the early 2000s the world-class point break that was Kirra point was literally buried.

The 'Bring Back Kirra' campaign led by the southern Gold Coast surfing community, which ran for a more than a decade, provides an example of the challenges which can occur within the policy stream. While the surfing community was clear that they wanted governments to intervene to 'Bring Back Kirra', there was a lack of consensus regarding what it was that governments should do to bring it back. This lack of consensus among stakeholders, regarding what would constitute a solution, provided all the excuse that policy makers needed to ignore the campaign through its initial phase.

The following quote from a Queensland State Government report on community attitudes to the Kirra Point Groyne highlights the perception by government that there wasn't a clearly preferred policy option among stakeholders.

> What to do with Kirra Point Groyne remains a significant and emotive issue, with some people advocating extending the existing groyne to its original length, while others believe that it should be removed or shortened further.
>
> (DEHP 2013)

The following quotes by three high-profile, former world champion Gold Coast surfers illustrate three different perspectives on Kirra Point Groyne.

> We just need to put the big groyne back to the dimensions it was before it was adjusted. That's not a theory, we have a 23-year data base that proves it.
>
> (Three-time world surfing champion Wayne Rabbit Bartholomew quoted by Feliu 2006, 6 July)

> The question is, will replacing the 30 metres removed from the artificial groyne in late 1996 help to bring back those famous Kirra Point Kegs?... There is no guarantee that replacing the front of Big Groyne can work like the old days.
>
> (1988 World Champion Longboarder Andrew McKinnon quoted by McKinnon 2011, 4 November)

> They are putting back the 30 metres and restoring the groyne to the original specifications. I've always said that the groyne is only part of the puzzle, there are things that need to happen before the wave will actually come back.
>
> (1990 World Champion Longboarder Wayne Deane quoted by Lockwood 2013, 21 August)

As the surfing community reached a consensus that the extension of Kirra Point Groyne to its original length was the preferred option, the issue attracted the significantly greater attention of policy makers. Through taking advantage of political opportunities in the lead up to election campaigns, the campaigners were successful in initially gaining support to profile the beach in 2009 and in finally gaining funding to extend the Kirra Point Groyne to its original length in 2014, almost a decade after the initial campaign began.

Gold Coast World Surfing Reserve

Another example of a solution in search of a problem and a political opportunity was the idea of establishing a Gold Coast World Surfing Reserve. A WSR is a programme and trademark of US non-profit organisation Save the Waves Coalition. The stated role of the WSR programme:

> proactively identifies, designates and preserves outstanding waves, surf zones and surrounding environments around the world.
>
> (Save the Waves 2016)

The idea of a Gold Coast world surfing reserve had been under discussion since the establishment of a Gold Coast National Surfing Reserve in 2012. NSR is a programme of a different Australian non-profit organisation, National Surfing Reserves.

> It means that the world and the nation recognises those three breaks here on the Gold Coast as being iconic … it also means that they have a level of protection so that in future those iconic surfing breaks are protected from, perhaps, inappropriate development.
>
> (Brad Farmer ABC Coast FM, Marshall 2012, 24 February)

Both NSR and WSR programmes claim to provide a form of recognition and promotion to 'iconic' surf breaks in addition to making claims of providing either protection or preservation. Neither of these non-regulatory/symbolic instruments provide further regulatory protection; however, they provide an important political foothold by which these conversations are able to develop. In the case of the Gold Coast, they also draw attention to surfing resources in a more systematic manner, rather than on a beach by beach or break by break scale.

The earlier section on crowding discusses how crowding is framed by policy systems, this recognition and promotion aspect of both programmes seems to align with the traditional policy system idea where attracting additional attention to surf breaks is a positive outcome. While this attention may be a positive outcome for local economic development for the Gold Coast, where surf breaks are among the most crowded in the world, there has been little attention given to resolving the impacts of this on local recreational surfing communities.

The protection or preservation aspect of non-government surfing reserve programmes is highly promoted by Save the Waves and NSR to local surfing communities; however, the actual mechanism for protection and preservation has been the subject of limited attention particularly given that the campaign to establish a Gold Coast WSR emerged in 2014 and was positioned as a direct response to a perceived threat of the construction of a cruise ship terminal at Kirra.

> [W]hen I heard about the proposed cruise ship terminal, my first thought was we're going to lose one of the greatest surf destinations on earth.
> (Mick Fanning quoted by Pawle 2014, January 10)

> [W]e need to take this to the next step and get this area approved as a World Surfing Reserve, we don't wanna come back every 10 or 15 years and protest against some surfing development that's gonna be detrimental to the area.
> (Former world champion and chair of Gold Coast World Surfing Reserve Committee Andrew McKinnon quoted by Smith 2015, 21 October)

The WSR was communicated as a higher level of 'protection' than the existing NSR, which was seen as having failed due to the Kirra cruise ship terminal proposal emerging within the area recognised as a NSR. However, the following statements from policy makers indicate that the proposal was far from being considered by government and may have been little more than a developer publicising an investment opportunity.

> It's nothing more than a sketch on the back of a napkin, despite what you've heard or read or even what the developers have said. There has been absolutely no proposal put to council or the state government. Practically speaking, there's nothing to protest against except a fanciful sketch.
> (Herman Vorster, then Media Advisor to the Gold Coast Mayor quoted by Workman 2014, 15 January)

> 'I haven't had any serious approaches about those off-the-beach type options', he said. 'I would have to say that I would be very sceptical that those types of options would ever receive the community support or government approval.'
> (Jeff Seeney, then Queensland Deputy Premier and Minster for Infrastructure and Development quoted by Pawle 2014, 10 January)

On the eve of a public rally organised by the advocates of the establishment of a Gold Coast WSR to oppose the Kirra cruise ship terminal, the Queensland Premier Campbell Newman released a statement that his government would not be supporting the development of a cruise ship terminal at Kirra.

> There would be no cruise ship terminal at Kirra Beach.... Our beaches are just too important – not just as a natural wonder or as a place for families and surfers – but also for the central role they play in Gold Coast tourism.
> (Campbell Newman, then Queensland Premier quoted by Ardern and Harbour 2014, 16 January)

While the threat to surf breaks from the Kirra cruise ship terminal development was without any enduring substance, the campaign to establish a Gold Coast WSR continued that spanned a significant geographical area (i.e. all of the Gold Coast initially and then the southern point breaks). Within the multiple streams context a surfing reserve is a policy solution that needs to align itself with a problem. From the surfing community the problem was clearly about protection of surfing resources illustrated by the following quote from Gold Coast professional surfer and former world champion.

> We need to protect our coastline from these type of developments and not find in 10–15 years that we are fighting all over again.
> (2012 WSL Champion Joel Parkinson quoted by Anon 2014, 4 February)

This alignment between the streams was contested by policy makers who disputed both the framing of the problem of coastal development threats to Surfing Capital and the solution as adding additional value to the current City of Gold Coast coastal management system.

> [W]hat aggravated this was this Kirra terminal thing. When people say we want to have this so we can have our beach protected ... I'm not against it as long as they can present to me that our community can benefit more than (the existing management plans and strategies) we already have.
> (Tom Tate quoted by Anon 2014, 4 February)

In response to the rejection of the problem definition by the policy system the WSR campaign reframed the problem away from protection and preservation (i.e. dealing with the critical demand and supply challenges outlined above) in communications with policy makers and towards recognition and promotion. The following quote shows that the WSR campaigners distanced themselves from seeking any controls on coastal management that would have supported achieving the intended protection and preservation objective

> WSR status will not impose costs or restrictions on Gold Coast City Council or State Government's management of ocean beaches. GCWSR will be honorific, thoughtful, and respectful of state and local authorities.
> (Report to City of Gold Coast by Gold Coast WSR nomination Campaign Team)

While the Gold Coast WSR nomination was eventually successful and was endorsed by Save the Waves in late 2015, the need to merge the solution with a problem recognised by policy makers has resulted in an outcome which shows limited immediate evidence of aligning with the problem perceived by the Gold Coast surfing community (i.e. active management of a modified beach and surf zone to benefit surfing and overcrowding) and which may in the end serve only to exacerbate existing crowding problems.

Politics stream

Within the multiple streams framework the politics stream refers to a wide range of factors that create the motivation and opportunity for policy makers to convert a policy concept into action. The politics of surfing on the Gold Coast should be understood as a counterbalance between the economic and social significance of what is simultaneously a sport, recreational activity and industry and the history of conflict between surfers and the City of Gold Coast bureaucracy which has led to a fractured relationship and ongoing tensions.

The Gold Coast is home to a number of professionalised surfing related organisations such as Surfing Australia and Surfing Queensland. In addition to these organisations the Gold Coast surfing industry has been estimated to be worth $3 billion each year (AECgroup 2009).

> 'We recognise and appreciate the fact that our city's enviable worldwide reputation is largely driven by the appeal of our surfing beaches so we're reminding the community that we're fully committed to protecting our beaches, supporting our surf industry and further developing our surf culture', he said. 'Surfing makes a vital contribution to our economy, contributing $3.3 billion annually and employing 21,000 people.'
>
> (Tom Tate quoted by City of Gold Coast 2014, 3 March)

> 'If a candidate doesn't have the surf community behind them on the southern Gold Coast they would struggle', he said. 'Candidates have been falling over each other to talk about issues like the Kirra and the CST down there. A lot of promises were made at the last state election.'
>
> (Greg Betts, then City of Gold Coast Councillor quoted by Simonot 2014,
> 31 December)

These quotes indicate that while there is recognition of the importance of surfing on the Gold Coast by policy makers, the relationship between surfers and the policy makers continues to be adversarial. The following quote identifies the disjuncture between the surfing community and policy makers.

> You can be the best surfer, diver, artist, I'm happy for you, but I will be listening to the people that matter.... It's more of an issue for high-profile people like the CEO of a cruise ship company ... they're the high-profile people I'm worried about.
>
> (Tom Tate quoted by Killoran 2012, 19 September)

This conflict between surfers and policy makers had deep roots with surfing groups opposing City of Gold Coast coastal management and development projects that date back to the 1960s.

Recent projects which have added to the tensions include the 1999 Narrowneck artificial reef and the ongoing Palm Beach shoreline protection project.

The Narrowneck project, an initiative of the City of Gold Coast, aims to reduce beach erosion through the construction of an artificial reef in the north of the city. This project is continually framed by surfers as an example of the incompetence of the City of Gold Coast. While the primary objective of the reef was erosion mitigation, at the time of its development the City promoted a secondary benefit of creating a new surf break, and this was exacerbated by claims from the reef designers (keen surfers themselves). The following quotes from Australian media illustrate the aims of the project and the perceptions of the surfers.

> This is the world's first multi purpose reef. It was set up for costal protection to protect the beaches of the gold coast and to improve the surfing. In the right conditions the break peels left and right, offering up two surfing waves from each swell.
>
> (Kerry Black Coastal Engineer and designer of the Narrowneck Artificial Reef quoted by Meerman 2008, 21 February)

> Mr (Wayne) Bartholomew (former surfing world champion) said the surf at the \$2.5 million artificial reef at Narrowneck was a 'hoax of a wave' because the sandbag reef was mainly designed to prevent beach erosion.
>
> (McElroy 2015, 12 June)

The perception of the failure of the City to deliver on the promoted surfing objectives of the Narrowneck artificial reef has had ongoing implications for the capacity to implement coastal management works on the Gold Coast. The Palm Beach Shoreline Project commenced in 1999 and similar to the Narrowneck project was an attempt by the city to reduce the vulnerability of Palm Beach to erosion. At the time Palm Beach was seen as an urgent priority. The project was opposed by the Palm Beach surfing community on the basis that the construction of a series of offshore submerged artificial reefs at Palm Beach would have negative consequences for the quality of beach breaks in the area. As a result of the opposition the Palm Beach shoreline project remains in the planning phase for now more than a decade. Despite numerous changes to the design of proposed reefs and the exploration of a number of alternatives such as artificial headlands, the City of Gold Coast bureaucrats have been unable to advance the project beyond planning due to community opposition.

The politics stream is about the motivation and opportunity for policy makers to convert a policy concept into action. The above analysis identified that economic value of surfing in the region, local electoral politics and the politics surrounding coastal engineering projects were the motivations and opportunities for policy development and implementation. How the three streams of problems, policy and politics came together as the Gold Coast Surf Management Plan is outlined below.

Merging the streams: Gold Coast Management Plan

In the previous sections we explored the disjunctures between the problems, policy and politics which have impacts on Surfing Capital. The discussion centred on the surfing spokespeople, surfing associations and surfing community actions and contrasted this with the coastal management activities of state and local government on the Gold Coast. This highlights significant friction between the surfing community and government, which produced a contested public policy landscape. This discussion describes the situation as of 2012. However, in late 2012 the surfing community, through its various representative organisations on the Gold Coast, instigated an approach to apply the multiple streams framework to reposition the interests of recreational surfers within the Gold Coast's coastal management policy system. The lead author of this chapter was instrumental, along with others from the surfing organisations, in forming the participatory, ground-up approach to the Gold Coast surfing social contract policy problem by drawing on the multiple streams theory.

The traditional relationship between recreational surfing and government had been constructed as protest by surfers against government initiatives. In a managed coastal process system such as the Gold Coast, by showing active resistance to coastal management or development, recreational surfers had positioned their interests in direct conflict to economic development. By altering their input to coastal management from 'resistance to change' to 'openness to opportunity' recreational surfers had the opportunity to reframe their relationship with government and develop the partnerships necessary to address threats to Surfing Capital.

In order to avoid the location-based politics of actions which may enhance Surfing Capital – be it avoiding development, dredging, sand pumping etc. – the solution was framed not as an initiative such as the development of a series of artificial reefs but as the development of a Surf Management Plan. Policy makers were well aware of the challenges in engaging with the surfing community, highlighted by delays or failure of coastal management and development projects. By linking the surf management plan to this challenge, presenting it as an opportunity to improve relations between surfers and the city hall, the problem and policy stream were merged.

The politics stream was shaped through use of existing institutional arrangements and the establishment of a coalition of surfing interests. City Councillor Greg Betts and member of the Gold Coast National Surfing Reserve Committee provided the formal mechanism to seek support from Gold Coast City Council as the community petition under the City of Gold Coast Subordinate Local Law No. 1.1 (Meetings) 2008. Under this law any petition presented to council by a councillor on behalf of the community with more than ten signatures should be considered for a review by council.

Surfrider Foundation Australia Gold Coast Branch invited surfing stakeholders to a briefing on the Surf Management Plan concept in 11 December 2012 at Kirra on the southern Gold Coast. Following a presentation by author Dan Ware on the potential to address threats to Surfing Capital in partnership rather than opposition to government through a surf management plan, the group agreed to sign a petition to be presented to Gold Coast City Council and to form a representative group Gold Coast Surf Council to work with government to better represent recreational surfing interests in coastal management. The founding members of GCSC included the members of the GCSC, representatives of Gold Coast boardriding clubs, Surfing Queensland, Surfrider Foundation Australia and National Surfing Reserves.

The following is an extract from the original petition from surfers to the City of Gold Coast to work collaboratively to develop the Gold Coast Surf Management Plan. The text highlights the language used to merge the various streams together.

> While Council, Industry and the Community all benefit greatly from the Gold Coast's surfing assets there are significant risks which if not proactively addressed will degrade these assets. Particular concerns of the undersigned include; crowding, increased vessel traffic, dredging, beach nourishment and development.
>
> There are major opportunities to maintain and enhance surfing assets through innovative design of coastal management initiatives as proven in other locations. By working cooperatively with surfing stakeholders to identify surfing assets and enhancement opportunities Council can prepare a surf management plan that will lead other surfing destinations and improve the surfing experience for community, tourists and industry.
>
> (Text from petition to City of Gold Coast – Ware 2012)

The City of Gold Coast engineering services committee considered the petition on the 14 February 2013 and adopted the recommendation that the draft Ocean Beaches Strategy include the development of a Surf Management Plan and a Commercial Activity Plan (City of Gold Coast 2014).

The terms of reference for the Surf Management Plan (City of Gold Coast 2014) required that the plan was to be developed and implemented through an ongoing consultative process which enables surfing representatives to have oversight on the implementation of the plan. The merging of problems, policy and politics in the policy landscape of maintaining Surfing Capital had been achieved.

At the time of publication the Surf Management Plan was scheduled for launch for 8 March 2016 following the successful endorsement by the City of Gold Coast in December 2015.

> Our beaches are for everyone and we want to ensure everyone can enjoy our city's prized natural attraction, this is a celebration of our City's Surf Management Plan and the soon-to-be World Surfing Reserve status.
>
> (Mayor Tom Tate, City of Gold Coast 2016, 9 February)

While it has been formally endorsed it remains to be seen what institutional impact the Surf Management Plan will have; however, this will be the subject of subsequent work by the authors.

Conclusion

The development of formal policy on the Gold Coast that recognised the threats to Surfing Capital is highly significant for a number of reasons. First, the process was successful in forming a coalition of disparate views and at key agenda-setting moments were critical in the formation of shared interests on policy formation. Second, the social contract of the Surf Management Plan was seen to be valid because of the ground-up, participatory process of development by the Gold Coast Surf Council. Third, the combination of Surf Management Plan and the Gold Coast World Surfing Reserves validates the social contract with surfers. Last, the accountability within the Surf Management Plan through participation by surfing groups increases the robustness of the social contract.

The previous adversarial relationship between surfers and government had negative consequences for the interests of both parties and the Surf Management Plan is a way to decrease these consequences. The multiple streams framework was applied on the ground to seek policy solutions to maintain Surfing Capital and the framework has been used here to analyse policy developments.

While the process has been positive, there are limitations – surfers as participants risk becoming co-opted by the process. Public conflict reduces trust between the bureaucracy and stakeholders so in seeking to build trust with government surfers may limit their public commentary, which may limit their power to negotiate a favourable outcome.

It is hoped that the surfing–government relationship will continue to improve within the now established policy framework, and new and ongoing issues can be addressed within this framework so that the interests of the surfing community can be valued along with other coastal users.

Note

1 Lazarow (2010) provides a detailed description of the modifications to Kirra Point.

References

AECgroup (2009) Surf Industry Review and Economic Contributions Assessment: Gold Coast City Council. Gold Coast, Gold Coast City Council: 52.

Anon (2012, 16 August) Cruise Ship Terminal Rally, Gold Coast Sun.

Anon (2014, 4 February) New Fight to Protect Aussie Beach Culture. www.news.com.au, accessed 10 February 2014.

Anon (2014, 19 November) Calls for Surf Police to Be Created to Tackle Rage at Gold Coast's Popular Breaks. Gold Coast Bulletin.

Ardern, L. and Harbour, J. (2014, 16 January) Premier Campbell Newman Kills Off Bilinga Cruise Ship Terminal Plan. Gold Coast Bulletin.

Berkman, K. (2012, 3 August) Critics Say Gold Coast Plan for Cruise Ships Won't Float. ABC News, 7.30 Report.

City of Gold Coast (2012, 2 November) Media Release. City of Gold Coast.

City of Gold Coast (2014) Engineering Services Committee Meeting Minutes, 14 February 2014, Gold Coast.

City of Gold Coast (2014, 3 March) Media Release. Gold Coast Mayor, City of Gold Coast.

City of Gold Coast (2016, 9 February) Media Release. Gold Coast Mayor, City of Gold Coast.

Cohen, M., March, J. and Olsen, J. (1972). A Garbage Can Model of Organisational Choice. Administrative Science Quarterly 17.1: 1–25.

Department of Environment and Heritage Protection (DEHP) (2013) Kirra Beach Restoration Project, Community Consultation Report. Queensland State Government, Brisbane.

Dredge, D., Hales, D. and Jamal, T. (2013) Community Case Study Research: Researcher Operacy, Embeddedness, and Making Research Matter. Tourism Analysis 18.1: 29–43.

Fanning, M. (2012, 18 September) www.mickfanning.com.au, accessed 18 September 2012.

Feliu, L. (2006, 6 July) Bring Back Kirra. Tweed Daily News.

Greenwood, E. (2014, 6 February) Kelly Slater Joins Pro Surfers Who Say Overcrowding at Snapper Rocks Is Making It Dangerous for Quiksilver Pro Surfers. Gold Coast Bulletin.

Killoran, M. (2012, 19 September) Surfing Champion Mick Fanning Slams Gold Coast Cruise Terminal Plans. Gold Coast Bulletin.

Kingdon, J.W. (2011) Agendas, Alternatives, And Public Policies. Boston, MA: Longman.

Lazarow, N. (2010) Managing and Valuing Coastal Resources: An Examination of the Importance of Local Knowledge and Surf Breaks to Coastal Communities. Fenner School of Environment and Society. Australian National University, Canberra. PhD.

Lazarow, N., Miller, M.L. and Blackwell, B. (2007) Dropping In: A Case Study Approach to Examine the Value of Recreational Surfing to Specific Locales. Shore and Beach 75.4: 21–31.

Lockwood, M. (2013, 21 August) Interview: Wayne Dean and the Rebuilding of Kirra Groyne. www.coastalwatch.com, accessed 15 September 2013.

McElroy, N. (2015, 12 June) Calls for Gold Coast Artificial Reef Intensify after Study Finds City Losing Millions as Surfers Dodge Crowded Breaks. Gold Coast Bulletin.

McKinnon, A. (2011, 4 November) Big Groyne Back on Agenda. Tweed Daily News.

Maritime Safety Queensland (2011) Currumbin Creek Bar Navigation Safety Report. Queensland State Government, Brisbane.

Marshall, C. (2012, 24 February) Local Beaches Made National Surfing Reserve. ABC Gold Coast FM.

Meerman, R. (2008, 21 February) Engineering the Perfect Wave. Catalyst ABC.

Moote, M.A., McClaren, M.P. and Chickering, D. (1997) Theory in Practice: Applying Participatory Democracy Theory to Public Land Planning. Environmental Management 21.1: 877–889.

Pawle, F. (2014, 10 January) Mick Fanning Joins Protest against Kirra Terminal, but Govt Is 'Sceptical'. The Australian.

Quinlivan, J. (2014, 10 December) Wayne Deane, Surfboard Shaper, Coolangatta. The Weekend Edition Gold Coast.

Rockloff, S.F. and Lockie, S. (2004) Participatory Tools for Coastal Zone Management: Use of Stakeholder Analysis and Social Mapping in Australia. Journal of Coastal Conservation 10.1: 81–92.

Save the Waves (2016) Protecting the Places You Love. www.savethewaves.org, accessed 15 March 2016.

Simonot, S. (2014, 31 December) Gold Coast Surfing Community to Push for Political Recognition of City's Second-Biggest Industry. Gold Coast Bulletin.

Slater, K. (2012, 21 September) www.twitter.com @kelleyslater.

Smith, J. (2015, 21 October) What the Gold Coast as a World Surfing Reserve Really Means. Stab Magazine.

Stuckey, J. (2013, 28 February) Media Release. Member for Currumbin Queensland State Government.

The Economist (2012, 17 March) Beach Rush: Surfers Hate Crowds and Need More Waves. Good News for Africa.

Vanclay, F. (2012) The Potential Application of Social Impact Assessment in Integrated Coastal Zone Management. Ocean & Coastal Management 68: 149–156.

Ware, D. (2012) Surf Management Plan Petition to City of Gold Coast. Unpublished.

Workman, A. (2014, 15 January) Save Kirra Now. Tracks Magazine. www.tracksmag.com, accessed 20 January 2014.

7 Surfers and public sphere protest

Protecting surfing environments

Rob Hales, Dan Ware and Neil Lazarow

Introduction

The counterculture of surfing developed with the growing popularity of surfing in the 1950s and 1960s and was in part a rejection of the dominant values associated with capitalism and materialism at that time, alongside a collective desire to break away from the cultural norms associated with the beach that had been dominated by the Surf Lifesaving movement in countries such as Australia and the USA (Booth, 2001; Jaggard, 1997; Pearson, 1979). Most surfers of that era would not have expressed it in those terms, more likely preferring to describe their approach as anti-authoritarian and non-conformist. The ironic feature of the counterculture of that time (and this may still be relevant today) was that the material benefits of economic growth (time and money) in the 1950s and 1960s created the very conditions for the counterculture to develop. Since then surfing culture has changed, but despite the mainstreaming of surfing culture and the creation of a surfing industry, the core 'anti-establishment' projection of surfing remain and keep alive the spirit of a counterculture. The feeling of freedom on a wave, the connection with elemental forces of nature, the social experience of surfing and physically immersing one's self in the ocean, is a potent cocktail that creates a strong emotional attachment to the places and people of surfing (Lazarow, 2010; Preston-Whyte, 2002). When these places are threatened, people respond with protest and campaigns to protect surfing environments. In this chapter we argue that surfing protest activities perform an important coalescing function that brings together the often fragmented elements of the broader surfing 'community'. These protest activities are a form of democratic response towards political and planning decisions that pose direct threats to the resources and values that these communities value – in other words, Surfing Capital (see Lazarow, 2010: 18). Surfing Capital refers to four factors which shape the surfing experience as (1) the physical features of and surfer's awareness of, the quality of the waves for surfing, (2) the frequency of quality waves, (3) the coastal and marine environment and (4) socio-cultural issues that are associated with coastal places.

This chapter examines the significant protest activities and public sphere campaigns surrounding resistance of surfers to the development perceived by surfers as inappropriate within the surfing environment and actions to protect these resources. Historically, surfers have resisted development in order to maintain wave breaks as common property, ensure beach access to the coastline and limit the impact of land-based activities that effect the beach and surf break environment. They have also protested against activities that impact on the marine environment. Whilst many of these actions can be considered self-motivated, the public good features of such actions are substantial. This chapter argues that the form of protest used by surfers creates, maintains and symbolises the common property features of surfing environments and through these very actions, surfers create a legitimate platform to participate in decision making.

The chapter is organised into three sections. First, we define the surfing public sphere, protest and the enclosure and externalisation of the commons. Next, we identify types of protest events in which surfers have engaged. This is followed by a case study of public sphere action where we examine an ill-fated cruise ship development proposal on the Gold Coast, Australia. This chapter makes a contribution to sustainability and surfing knowledge through providing evidence that surfer's resistance to development is made possible by the type of public sphere action with surfing environments and thus surfing resistance can be seen as more than self-centred expressions to maintain lifestyles and hedonistic pleasure. It also identifies the growing power of surfing communities in public sphere disputes.

The surfing public sphere, protest

The origins of the term public sphere describe the rise of the bourgeois culture where public debate through the civil society contributed and influenced the decisions of government (Habermas, 1991). The public sphere is not a singular place but rather there are multiple spaces where the public can variously atempt to express matters that they consider important to society and through this, influence both the public perception as well as decision making in the political sphere (Fraser, 1990). The public sphere is often conflated with the media but is not limited to the media only. The surfing public sphere can have a physical location and applied to surfing can be the space of the waves. The joining of surfers for a paddle out in the public space of the waves for the purpose of influencing cultural values and political decisions can be considered a public sphere, especially if there is media coverage of the event.

In recent times, the development of information communication technology has expanded the public sphere. Contemporary media communications facilitate wider community engagement. The far-reaching capacity of Facebook, for example, can be considered part of the public sphere because media (and politicians) monitor social media activity as a way to gauge the sentiment of the public on a particular topic.

From the outset we should mention here the scope of this chapter predominantly examines the public sphere in developed countries because the concept of the surfing common is linked to how local communities (government) enact their land/sea tenure systems of 'ownership' (Rider, 1998). Not all surf breaks are common property. For example, Fiji had a customary land tenure system in which particular people and families own, or are custodians of, certain sea areas (Ponting and O'Brien, 2013).

Protest also needs to be defined for the purpose of the chapter. Protest is defined as a direct action undertaken by surfers to resist development that degrades Surfing Capital and is an action within a campaign advocating for the maintenance of Surfing Capital. Protest events where media is involved that promote the cause of surfers or public campaigns elevate the protest into the realm of a public sphere. Similarly, a large mass protest of surfers at a rally is also the public sphere irrespective of whether media is present or not.

Enclosure of the commons and protest

Increasingly, there is recognition of the negative impacts of continued economic development which encloses or limits access to public space (commons) in the pursuit of continued economic growth (Jeffrey *et al.*, 2012; Springer, 2011). One of the drivers of economic growth is the process of including previously unallocated public land/sea (commons) within capitalist production systems (Monbiot, 1994). This process is called enclosure and the feature that enables this process is the purposeful public sphere manipulation of the value of that space by proponents so that there is little or no value of that space to society for any other purpose except for the particular outcomes of economic development (Sevilla-Buitrago, 2015).

This is relevant to surfing because the process of enclosure used by the proponents of coastal developments routinely devalues surfing so as to position economic development as the only valued outcome for that coastal environment. This process has been increasingly challenged by surfers. Surfers, through their actions of surfing on the waves, have increasingly commodified the surfing common and thus, its value to society and the economy has also increased. Thus surfing through using the common, albeit in a commodified way, maintain the surfing environment in a sustainable way as opposed to coastal development which severely alters the natural environment. It should also be noted here we are not arguing that surfers who protest to maintain the common property features of coastal environments (and the quality of those environments) are in some way morally superior to others. In fact, Lazarow and Olive in this volume report that many surfers believe other types of outdoor recreation user groups have a lighter environmental footprint than surfers. The moral terrain of environmentalism is complex. Surfing complicates this further through our oft-colonial approach to communities and resources in less-developed regions of the planet; and this is reflected at an individual (i.e. the recreational surfer) level and also through the lens of the surf industry (Ponting, 2001; Ponting *et al.*, 2005).

From a resistance to enclosure perspective, surfer's resistance to development not only is seen as a reaction to preserve nature and or maintain surfing environments but also positions surfers as performatively creating and maintaining public interest values of coastal environments through protest in that space. By protesting in the public space of beaches and waves, the very act of protesting reaffirms public interest values of the place and common pool resources. Surfers who protest implicitly know this, so it is no coincidence that one of the most common places of protest for surfers is the waves and beaches of surfing. Protest paddle outs and rallies on beaches performatively create and claim the space of the waves and beaches as a common and through media coverage of such events creates the space of the public sphere as the waves also. This is a powerful combination in terms of political advocacy for surfing as it disrupts the devaluing process or proponents who attempt to enclose the commons.

Not all protests that occur in the surfing public sphere are about protecting the commons from enclosure (protecting waves, accces and marine environments). The other issue that surfers are concerned about is that the surfing common suffers from the externalities of industrial development. That is, the surfing environment used as either an overt or inadvertent dumping ground for communities linked to the coast. Protest events outlined below highlight the types of surfing protest.

Types of surfing protest and organisations

The types of protest that surfers engage with can be categorised in four ways and reflect the location and issues at hand. The four ways identified in this chapter are:

1   protesting against developments that impact on waves;
2   protesting against developments near beaches including water quality impacts;
3   protesting against the loss of access to surfing beaches; and
4   surfer protests linked to other environmental and social issues.

These four ways of protest link with the four primary elements of Surfing Capital identified by Lazarow *et al.* (2007), which are: wave quality, wave frequency, environmental matters and socio-cultural issues. When surfers identify elements of Surfing Capital are significantly threatened then protest (and advocacy) is the likely result. Three tables are presented to show three types of protest: protest over development impacts on waves, protest over developments near beaches including impacts on water quality and protest over issues of access to waves. Surfer protests linked to other environmental and social issues will be discussed following this. The tables are simply indicative of where surfing communities have engaged in direct protests as part of a broader advocacy campaign.[1] An indicative list of examples of where direct action protests have been used as part of more substantial public advocacy campaigns is provided in Table 7.1. The documentation of these protests below in such an abbreviated way underrepresent the years of effort many people have invested in the campaigns associated with the protests documented here.

Table 7.1 Protests against developments impacting on waves

Year	Location	Direct action protest details
2015	La Pampilla beach, Peru	Surfers and police clash over road expansion project which will impact on the quality of a point break
2014–2015	Montauk, NY, USA	Proposed geotextile seawall at Montauk Beach halted. Direct action includes civil disobedience and paddle outs
2014	Kirra, Gold Coast, Australia	Proposal for cruise ship terminal development. The project was scrapped as a result of a public protest rally and (planned) paddle out. (Surf conditions stopped planned paddle out)
2004–2013	Mallacoota, Victoria, Australia	Proposed harbour development across surf-break. Numerous protests during the campaign
2011	Aramoana, Dunedin, New Zealand	Paddle out as protest action over dredge spoil impacting waves and beach quality
2007	Asturia, Spain	Surfers hold protest rally over Rodiles dredging project
2006	São Miguel in the Azores, Portugal	A series of protests over a number of port developments in this region
2005	Dana Point, CA, USA	Protest against seawall proposal in 2005 and success for surfers in 2012 (there was also access issues and protest in 2009)
2004	La Herradura, Peru	Developers agreed to scrap a marina proposal
2001	Lugar de baixo, Madeira, Portugal	Public rally as part of campaign to halt jetty construction that had adverse impact waves

Across cultures and countries, the use of direct action protest has been a common tactic deployed by surfing communities and activists to draw attention to their issues. The spectacle of paddle outs and the use of surfboards and wetsuits as props have been particularly effective in gaining media attention, especially away from the beach (e.g. in city centres or outside parliamentary buildings). Increasingly activists understand the linkage between political opportunities of staged contentious performance and media uptake of the logic of their protesters' campaign (Cammaerts, 2012).

Successful protesting needs the effective organisation of people. A number of surfing non-governmental organisations (NGO)/community groups found their genesis in the continued interest from surfers in the wake of early advocacy campaigns and protests, for example, Surfrider Foundation and Save the Waves Coalition. These organisations have developed in response to localised protest events and campaigns and now mobilise behind local surfing communities in campaigns to promote the effective management of wave resources as part of a broader ecologically sustainable development charter. The World Surfing Reserve campaign is a reaction to proactively circumvent inappropriate development of the surfing environment.

The UK-based Surfers Against Sewerage is another successful surfing NGO that campaigns against coastal development impacting on waves. However, the original protests which kick-started the organisation targeted the water quality impacts of nearshore sewerage outfall near the coastal towns of St Agnes and Porthtowan in Cornwall, England. Table 7.2 provides an indicative set of examples of where surfing communities have protested against developments near beaches including water quality impacts.

The key theme running through this type of protest is that surfers experience first-hand the impacts of industrial development on their bodies. As a the first protest by surfers against sewerage in Cornwall, England attests, snorkelling gear may be needed to protect one's self when surfing in those places. To go surfing in the sea where there is direct sewerage outfall should necessitate protective clothing and demonstrating this through a protest event makes for a 'potent' media spectacle.

Protests against developments impacting on water quality are not only about maintaining surfing amenity but also have a greater purpose for protecting the nature of the surfing environment as part of the common pool of resources. Similar to the protests by surfers who opposed developments impacting directly on waves, protesting using paddle outs clearly symbolises that the public sphere of surfing is actually the waves themselves. The third type of protest that we examine relates to accessing the surfing common (see Table 7.3).

Access to the common property of waves is another issue that prompts surfers to protest to maintain their access rights. Access is critical to common property. Many countries have a system of land tenure where beaches can be private property and as such this can create access problems for surfers. Where beaches are common property and owned by the state there can still be access issues in that road and pathways access can be limited by private property adjoining the beach, for example, The Ranch in California.

The fourth protest type identified was that surfers link their protest with other environmental and social issues. Issues such as climate change, dolphin and whale harvesting, plastic pollution are issues that resonate strongly with surfing interests groups. Although this is difficult to evidence, in our opinion the level of protest by surfers is somewhat less than the other three types but in terms of campaigns these issues are receiving growing attention from surfing organisations in recent years.

Table 7.2 Protests against developments near beaches including water quality impact

Year	Location	Direct action protest details
2011	Pavones, Costa Rica	Tuna fish farm pollution protest
2009	Long Beach, CA, USA	Surfers protest against liquefied natural gas plant near long beach
2007	Hawai'i, USA	Surfing demonstrators protest Hawai'i ferry
2008–2010	Jeffrey's Bay, South Africa	Nuclear power plant construction proposed – protest with other groups
1990	England	Surfers against sewerage protest

Table 7.3 Protests against loss of access to surfing beaches

Year	Location	Direct action protest details
2013	Thanburudhoo, Maldives	Privatisation threat to surf breaks on islands instigates protests from locals and expatriates. Only people staying on the islands would be able to surf.
2012	Martin's Beach, CA, USA	Surfrider Foundation organises protests over access to beach
2009	Broadbench, England	350 surfers engage in paddle out to protest closure of beach access
2006	Trestles, CA, USA	Toll road proposed, which would restrict beach access and impact sediment and substrate supply to the nearshore, and subsequent protest organised
2006	Playa Encuentro, Dominican Republic	Public campaign to stop privatisation of beach and access problems
1997–2005	Asbury Park, NJ, USA	The campaign to permit surfing at beaches and remove beach entry fee was successful. Significant advocacy but no direct action protests

Gold Coast surfing protests

A number of changes in the world of surfing and the societies in which surfing occurs have produced a defined surfing public sphere in various countries, particularly in places that are deemed surfing hotspots. These changes contribute to the growing force of surfing and give weight to protest actions that were once considered protest actions of people on the fringe of societies. A case of surfers protesting against a cruise ship terminal on the Gold Coast, Queensland is explored below to examine the features that now make surfing more powerful in the public sphere.

The Gold Coast is home to a number of world-class point break waves within its 40 km of urbanised open beaches, all of which produce surfable waves. It is conservatively estimated that of the 12–14 per cent of Gold Coast residents (i.e. approximately 65,000) participate in surfing. Additionally, it is estimated that around 30 per cent of surfing effort on the Gold Coast can be attributed to non-residents (Lazarow, 2010). Surfing is one of the iconic tourism images promoted by the local tourism industry destination management body: Tourism Gold Coast. Surfing is not only important for local residents who surf but the local tourism industry, which relies on positive images of the Gold Coast as a surfing destination.

Local residents have been actively campaigning to protect surfing breaks and the surfing environment since the early 1970s. Table 7.4 describes the most significant campaigns to protect Surfing Capital on the Gold Coast.

The one issue that dominates the protest public sphere on the Gold Coast is cruise ship port development. Over the past 30 years, there have been repeated attempts to build cruise ship ports at various headlands and river entrances on

Table 7.4 List of issues where surfers have protested against development impacting on surfing environments

Year	List of surfing protests on the Gold Coast, Australia
1972	Original Kirra Groyne was protested by Bill Stafford the head of Surfing Queensland at the Coolangatta Chamber of Commerce Meeting
1973	Protests by Surfing Queensland against the Currumbin Creek rock walls
1980s	Casino and cruise ship terminal for Currumbin Creek
1980s	Marina proposal at Kirra Point
2004–2005	Palm Beach Protection Strategy, inclusive of artificial reefs
2005–2006	The cruise ship terminal at The Spit in 2005
2014	Kirra cruise ship terminal
2015	Cruise ship terminal, The Spit

the Gold Coast, most recently in 2014 and 2015. If approved both proposals were perceived to negatively impact on surfing conditions in those areas, amongst an array of more and less significant impacts. Of course, there is always the likelihood that new or augmented surfing conditions might be a by-product of the proposed developments. But at what price the risk?[2] The purpose of the following analysis is to highlight which features of the public sphere actions were important in the emerging legitimacy of surfers protest actions.

The first feature that was important was the organising capacity of surfing groups and related groups to mobilise large numbers of people to protest. Through surfers' online networks the groups managed to organise over 2,500 people to protest on 19 January 2014 in opposition to the proposal at Kirra. The Save Our Southern Beaches Alliance used Facebook pages to link with other groups and individuals to publicise the event. Given the history of similar proposals and the ongoing 'Bring Back Kirra' campaign, there was already a network of people who could organise and mobilise the protest.

The second feature identified was the growing economic importance of surfing, and the economic values of waves have meant that surfing had gained standing in government and bureaucratic decision-making processes. A number of reports had identified the economic value of surfing on the Gold Coast (AECgroup, 2009; Lazarow, 2009; Lazarow *et al.*, 2008b). There were also a growing number of surf economic studies examining total economic value (see Costanza *et al.*, 1997 for a description of TEV; Lazarow *et al.*, 2013) and putting a price on surfing (Bicudo and Horta, 2009; Chapman and Hanemann, 2001; Durham and Driscoll, 2010; Lazarow, 2010; Lazarow *et al.*, 2007, 2008a; Nelsen *et al.*, 2007; Raybould and Lazarow, 2009). A central facet of many of the surf economic studies has been a view that monetising the surfing experience provides the surfing community with an important weapon alongside the cultural and physical values, to combat the proponents of inappropriate development. The use of economic arguments, one might argue, has somewhat entrenched the commodification of wave resources, but in reality, this has simply drawn out what many have known for some time – that is, that surfing resources underpin the economic development of many towns, cities and regions, and, of course, the global surfing industry. The growing research in this area gives weight to the economic argument.

The third feature that led to the effective public sphere action is concerned with the relationship between surfing culture and protest. Surfing culture has media appeal and, therefore, surfing protest issues can be readily taken up by media outlets because the protesters have been legitimised in the mainstream culture in Australia. Notable professional surfers who have had media attention because of their success in the World Surfing League added their voice to the protests. For example, Mick Fanning made a series of strong statements against the Kirra proposal and also did not shy away from defending himself against the remarks of the mayor of the Gold Coast over Mick's involvement in the campaign. Also not always apparent in the public eye is that people who are surfers are now also part of the political establishment and also part of the administrative sphere of government. Thus, there are sympathetic allies involved in public policy decision-making processes.

The last important feature we have identified concerns government elections and surfing. As a result of the growing popularity of surfing, surfers represent an important interest group from an electioneering perspective of political candidates. At the time of the 2014 Kirra protest, there was a looming state election (there are three levels of government in Australia – local, state and federal) and the political parties attempted to align themselves with surfers and their interests in order to secure surfers' votes. This theme has repeated itself over the past few state elections, even to the point where Conservative Party supporters donned 'Save Kirra' T-shirts at the manned election stations to hand out how-to-vote cards in the 2009 elections. The electoral seat where the cruise ship was proposed was a safe seat held by a member of the ruling Conservative Party prior to the 2014 election. The decision not to proceed with the terminal at Kirra was decided by the state government and the decision could be seen as a way of ensuring the safe seat would be secured for the incumbent Conservative Party during the next election. Interestingly, a decision by the then Labor government in 2006 to overturn their own earlier proposal for a cruise ship terminal in the Gold Coast Seaway did not bring them much closer to winning the seat.

Concluding comments

This chapter has outlined the ways in which surfers protest, and argued that the form of protest (and advocacy) used by surfers creates, maintains and symbolises the common property features of surfing environments and through these very actions, surfers create a legitimate platform to participate in decision making. Most surfing protest and public campaigns can be viewed as resistance to the enclosure of the commons by inappropriate development. This feature of protest in the public sphere was highlighted in a short case study of protest against a cruise ship terminal on the Gold Coast. Surfing protests also centred on externalities of industrial development and conservation of the marine environment. The concept of Surfing Capital helped explain why surfers would protest against a certain issue.

Our analysis from examining the case study and the types of recent advocacy and protest events outlined above is that the success of public sphere action is due to a number of factors. First, the organising capacity of surfing groups facilitated through surfers' online networks has increased in sophistication. Second, the growing economic importance of surfing and the values of waves have meant that surfing has gained standing in bureaucratic decision-making processes. Third, surfing culture has media appeal and, therefore, surfing protest issues are taken up by media outlets because the protests are from a group that have been legitimised in the mainstream culture. Last, the growing popularity of surfing means that surfers count in local democratic elections and thus surfing issues are on the political agenda. These issues resonate strongly with earlier work by Lazarow (2010)

The features of surfing protest highlight the growing power and influence of surfing activism in the public policy decisions affecting our coasts. The development of coasts throughout the world will most likely continue but from a surfing activism perspective, the likelihood of successful campaigns to preserve surfing and coastal environmental assets and values is higher given the rise of surfing advocacy.

Notes

1 For a geographical overview of recent protests and threatened waves please refer to organisations such as Surfrider Foundation and Save the Waves for information about the location and details of campaigns.
2 For an overview the history of the dispute please see the relevant websites of the interest groups: Save Our Sothern Beaches and Save Our Spit.

References

AECgroup. (2009). *Surf Industry Review and Economic Contributions Assessment: Gold Coast City Council*. Gold Coast: Gold Coast City Council.

Bicudo, P. and Horta, A. (2009). Integrating Surfing in the Socio-Economic and Morphology and Coastal Dynamic Impacts of the Environmental Evaluation of Coastal Projects. *Journal of Coastal Research, SI56*, 1115–1119.

Booth, D. (2001). *Australian Beach Cultures: The History of Sun, Sand and Surf*. London: Frank Cass.

Cammaerts, B. (2012). Protest Logics and the Mediation Opportunity Structure. *European Journal of Communication, 27*(2), 117–134.

Chapman, D. J. and Hanemann, W. M. (2001). Environmental Damages in Court. In Heyes, A (ed.), *The Law and Economics of the Environment* (pp. 319–367). Northampton, MA: Edward Elgar.

Costanza, R., D'Arge, R., De Groot, R., Farber, S., Grasso, M., Hannon, B., Limburg, K., Naeem, S., O'Neill, R. V., Paruelo, J., Raskin, R. G., Sutton P., and Van Den Belt, M. (1997). The Value of the World's Ecosystem Services and Natural Capital. *Nature, 387*, 253–260.

Durham, W. and Driscoll, L. (2010). The Value of a Wave: An Analysis of the Mavericks Wave from an Ecotourism Perspective, Half Moon Bay, California. California: Center for Responsible Travel and Save the Waves Coalition.

Fraser, N. (1990). Rethinking the Public Sphere: A Contribution to the Critique of Actually Existing Democracy. *Social Text, (25/26)*, 56–80.

Habermas, J. (1991). *The Structural Transformation of the Public Sphere: An Inquiry into a Category of Bourgeois Society*. Cambridge, MA: MIT Press.

Jaggard, E. (1997). Chameleons in the Surf. *Australian Journal of Studies, 53*, 183–190.

Jeffrey, A., McFarlane, C. and Vasudevan, A. (2012). Rethinking Enclosure: Space, Subjectivity and the Commons. *Antipode, 44*(4), 1247–1267. doi: 10.1111/j.1467-8330.2011.00954.x.

Lazarow, N. (2009). Using Observed Market Expenditure to Estimate the Economic Impact of Recreational Surfing to the Gold Coast, Australia. *Journal of Coastal Research, Special Issue 56: Proceedings of the 10th International Coastal Symposium*, 1130–1134.

Lazarow, N. (2010). *Managing and Valuing Coastal Resources: An Examination of the Importance of Local Knowledge and Surf Breaks to Coastal Communities*. PhD, Australian National University, Canberra.

Lazarow, N., Miller, M. and Blackwell, B. (2007). Dropping In: A Case Study Approach to Examine the Value of Recreational Surfing to Specific Locales. *Shore and Beach, 75*(4), 21–31.

Lazarow, N., Miller, M. L. and Blackwell, B. (2008a). The Value of Recreational Surfing to Society. *Tourism in Marine Environments, 5*(2–3), 145–158.

Lazarow, N., Tomlinson, R., Pointeau, R., Strauss, D., Noriega, R., Kirkpatrick, S. and Stuart, G. (2008b). *Gold Coast Shoreline Management Plan. Volume 1: Executive Summary and Littoral Review Part A*. Gold Coast: Griffith Centre for Coastal Management.

Lazarow, N., Raybould, M. and Anning, D. (2013). Beach, Sun and Surf Tourism. In Tisdell, C. (ed.), *The Handbook of Tourism Economics: Analysis, New Applications and Case Studies* (ch. 17). Singapore: World Scientific Publishing Company.

Monbiot, G. (1994). The Tragedy of Enclosure. *Scientific American, 270*(1), 159.

Nelsen, C., Lazarow N., Bernal, M., Murphy, M. and Pijoan, P. (2007). *The Socioeconomics and Management of Surfing Areas: International Case Studies from Mexico, Spain, California and Australia*. Paper presented at the Coastal Society: 21st International Conference, Redondo Beach. www.surfrider.org/surfecon/tcs_surfecon_session.pdf, accessed 1 August 2016.

Pearson, K. (1979). *Surfing Subcultures of Australia and New Zealand*. Brisbane: University of Queensland Press.

Ponting, J. (2001). *Managing the Mentawis: An Examination of Sustainable Tourism and the Surfing Tourism Industry in the Mentawi Archipelgao, Indonesia*. Master's Thesis, University of Technology, Sydney.

Ponting, J. and O'Brien, D. (2013). Liberalizing Nirvana: An Analysis of the Consequences of Common Pool Resource Deregulation for the Sustainability of Fiji's Surf Tourism Industry. *Journal of Sustainable Tourism* (ahead-of-print): 1–19.

Ponting, J., McDonald M. and Wearing, S. (2005). De-Constructing Wonderland: Surfing Tourism in the Mentawi Islands, Indonesia. *Society and Leisure, 28*(1), 141–162.

Preston-Whyte, R. (2002). Constructions of Surfing Space at Durban, South Africa. *Tourism Geographies, 4*(3), 307–328.

Raybould, M. and Lazarow, N. (2009). Economic and Social Values of Beach Recreation on the Gold Coast. CRCST Project #100054 Technical Report. Gold Coast: Griffith University & CRC for Sustainable Tourism.

Rider, R. (1998). Hangin'ten: The Common-Pool Resource Problem of Surfing. *Public Choice, 97*(1–2), 49–64.

Sevilla-Buitrago, A. (2015). Capitalist Formations of Enclosure: Space and the Extinction of the Commons. *Antipode, 47*(4), 999–1020.

Springer, S. (2011). Public Space as Emancipation: Meditations on Anarchism, Radical Democracy, Neoliberalism and Violence. *Antipode, 43*(2), 525–562.

8 The non-market value of surfing and its body policy implications

Jason Scorse and Trent Hodges

There is a small, but growing work related to the economics of surfing inform-ally known as 'surfonomics'. This term encompasses a vast array of methodolo-gies and techniques designed to capture the economic impact and significance of the sport of surfing. Most economic studies of surfing have examined the direct expenditures that surfing produces for the local economy – e.g. hotel bookings, restaurants, surf lessons, etc. This technique produces a direct market value and is useful because it captures the surfer as a tourist participating in the local economy through their daily purchases.

Some of these studies have demonstrated that in many areas surfing adds mil-lions of dollars to the local economy. In Mundaka, Spain, Murphy and Bernal (2008) designed an online survey to capture the direct expenditures of tourist surfers and found that the average surfer spends about $120 a day, which when combined with visitation rates totalled between $1 and $5 million in spending per year. Using a survey-based methodology, Nelson *et al.* (2007) calculated average surfer expenditure per trip of $40.20 with a total annual economic con-tribution between $8 and $13 million in direct spending at the surf spot Trestles in Southern California. The basic expenditure model has become a well-established and effective method for capturing local economic impacts of surfing. The international non-profit organization, Save the Waves Coalition, based in Santa Cruz, California has a programme dedicated to expenditure ana-lysis in specific surf areas where the organization works on promoting conserva-tion. A Save the Waves study in Pichilemu, Chile found that surfers bring in between $2 to $8 million in direct expenditures, and forthcoming studies in Huanchaco, Peru and Bahia de Todos Santos are using this method to capture this direct market value of surfing as well.

Other studies have used a similar methodology but extended the findings to a much larger region. Neil Lazarow observed market expenditures to estimate the value of recreational surfing on the Gold Coast of Australia, an area that receives over nine million visits a year. He reported that the annual expenditure of recrea-tional surfers in 2007 was estimated to range from AU$126–233 or about US$89–164 million annually (Lazarow 2009).

There are many other sources of direct market contributions from surfing such as international contests, webcam businesses, and global merchandise. The

North Shore of Oahu hosts the Vans Triple Crown of Surfing, one of the most prestigious and well-known events in professional surfing. In 2010, an estimated 23,195 physical spectators spent an average of $178.13 per day during the 12 days of competition. In 2011, a detailed analysis carried out under the direction of Lenard Huff at Brigham Young University-Hawai'i shows that the event generated $20.9 million in direct expenditures from this single event (Reed 2011).

Direct expenditure analysis is a firmly established method to estimate the direct market economic values associated with surf tourism. However, surfing also produces significant non-market economic values that are not captured by any kind of expenditure studies. These non-market values fall into one of three categories: (1) non-use value, which is non-consumptive or indirect, and (2) consumer surplus and (3) capitalized real-estate value, both of which are consumptive and direct.[1]

To date, no study has yet been published on the non-use value of surfing. Such research would include the existence value people hold from simply knowing that surfing resources are being protected (even if they don't use them directly), the option value of preserving a surf location for future use, or the bequest value of being able to pass surfing resources down to future generations. There are likely many surfers who value waves they have not yet surfed, or may never surf, but for whom preserving waves still holds economic value. The absolute amount of economic value per individual might be small but with tens of millions of surfers worldwide (and aspiring surfers) the sum could be quite large.

Because non-use values do not leave 'behavioural footprints' (i.e. they only exist in people's minds), conducting non-use value research requires survey work, typically using the Contingent Valuation Method (CVM). By posing hypothetical situations that can elicit true willingness-to-pay responses for non-use values, CVM can potentially capture values that we don't observe but which are real. However, to do CVM well requires a good deal of time and money, and many CVM studies (even many published in peer-reviewed journals) do not meet the protocols established by the 1993 'Report of the NOAA Panel on Contingent Valuation' that set the standards.[2] Furthermore, Chapman and Hanemann (2001) have argued that current studies using contingent valuation to estimate values for California beach visits may not be reliable because the surveys are not site specific and fail to account for variation in travel cost to beaches throughout the state.

Non-use values are controversial because measuring them is so difficult. In the economics literature, there are long-standing feuds between economists with radically different views on the validity of using CVM to estimate non-use values, and whether these values should be used in public policy. However, resource economists have generally accepted CVM as a reliable method for environmental valuation and many argue it has proven no less reliable than behavioural methods of measurement (Haab and McConnell 2002). Many national courts (including in the US) have sided with those who believe there is a place for non-use values in the policy process, and these non-use values are also often consequential in the case of natural resource damage claims (Carson *et al.* 2003). For example, after an oil spill, companies are often held liable for the damage to fisheries and tourist industries, but in many jurisdictions they can also be held accountable for the lost non-use value. In fact, Carson *et al.* (2003) note that the US Oil Pollution Control Act of 1990 came down on the side of including non-use values in assessing damages for oil spills. This can greatly increase the total damage claims levied by the state. If a pristine marine environment is despoiled (e.g. Prince William Sound in Alaska after the Exxon Valdez oil spill of 1989), the public loses not only the direct economic benefits from the resources but whatever values were derived from existence, option, and bequest values. If a surf break were despoiled, the same would be true.

The most common type of non-market value studied in surfing is consumer surplus, and this is commonly estimated using the Travel Cost Method (TCM). TCM uses regression analysis based on data of individuals' distance travelled to a surf location, along with fuel and time costs, to create a demand curve for surfing; the number of trips is the dependent variable and the travel costs are the independent variables. Oftentimes, other independent variables are included in the analysis, such as the quality of the resource, and demographic variables such as age, income, and education.

Once this demand curve is estimated, the researcher(s) uses this equation to estimate the demand for the surf resource as the price hypothetically rises, to the point at which the surf location is so expensive that no one would choose to surf there. The area under the demand curve from the current demand (with no added cost) to the hypothetical 'choke price' at which surf participation is reduced to zero measures the consumer surplus in dollars. The consumer surplus represents the aggregate willingness-to-pay above and beyond what people are currently paying to get to the surf location. Think of it this way: every time people spend money to get to a surf spot, many of them would've likely spent a little more money if this was required (whether in fuel or time). The truth is that the amount of money people currently pay to go surfing is not necessarily their *maximum willingness-to-pay*. The difference between what they currently pay and the maximum willingness-to-pay is the consumer surplus, and this is what TCM allows researchers to estimate for specific surf locations.

As a complement to direct expenditure studies, TCM surfonomics reports have become a popular method of assessing the economic value of surfing. This methodology has a robust history as it relates to beach recreation and has since been used specifically for surfing. For example, Chapman and Hanneman (2001) estimated a consumer surplus value of $13 dollars per person at Huntington Beach for all recreation activities. Pendleton and Kildow (2006) used various TCM studies to estimate the total non-market value of beach recreation in California at more than $2 billion annually.

The first known TCM to estimate consumer surplus for surfing was completed by Charles Tilley, a Master's student at California State University Monterey Bay, who estimated the value of consumer surplus at Pleasure Point, California. He found through a TCM assessment of the famous surf break that surfers represent $8.4 million of consumer surplus (Tilley, 2001).

Chad Nelsen, currently the Executive Director of Surfrider, published his dissertation from UCLA in 2012, which includes a detailed single-site TCM approach for the famous surf spot known as Trestles in San Clemente, California. The estimated per visit consumer surplus value of $138 extrapolated to a range of visits per year resulted in $21 to $45 million dollars annually in total consumer surplus (Nelsen, 2012).

Surfing is also a spectator sport and much of its economic value can be attributed to those that go to certain places to watch the sport. A study in 2009 by Coffman and Burnett attempted to estimate the consumer surplus through a TCM analysis at the famous big wave Mavericks in Northern California. They found that an average visitor receives a benefit of $56.7 per trip, which amounts to a total economic benefit to the region of about $23.8 million (Coffman and Burnett, 2009).

Another way to think of consumer surplus in the context of surfing is that for any given surf spot there are people who live far away from the resource, and therefore must commute to it, and those who live close by and can either walk, ride a bike, or perhaps take a short car ride. Those who live closer on average receive higher consumer surplus because they have to expend less money and time (which is also money) to get to the resource, and therefore, the difference between what they currently expend and their maximum WTP is likely higher.

These TCM surf studies, such as Nelsen's Trestles study, have been done primarily in locations where the surf spots are frequented mostly by locals and other nationals who live within driving distance. This makes sense because the economics of remote surf destinations that cost thousands of dollars to get to and are used mostly as vacation destinations are quite different. The notion of consumer surplus doesn't make much sense for these destinations because travellers have many options to choose from when considering an expensive trip, and oftentimes the number of foreign visitors outnumbers the locals, thereby complicating the analysis.[3]

As Scorse *et al.* (2015) argue in their paper 'Impact of Surf Spots on Home Prices in Santa Cruz, CA', the issue of consumer surplus for surfing is further complicated by the fact that real estate prices are likely impacted by proximity to surf breaks. While people who live close to surf spots have much lower travel costs, they may pay higher home prices to live close to surf resources, and therefore, much of the consumer surplus would then be capitalized into real estate values.

The Hedonic Price Method (HPM) estimates the independent contribution of different home characteristics to the overall home value, and can be used to value backyards, ocean views, or even clean air, school quality, and proximity to natural resources such as surf breaks. Analysing how real estate values are impacted by proximity to surf breaks provides a non-market estimate of the consumptive non-market value of surfing that is *embedded in the market prices of real estate*.

But estimating the impact of surfing on home prices is significantly more difficult than estimating consumer surplus using TCM. While the statistical techniques are not necessarily more complicated, it is often hard to find coastal regions with real estate adjacent to beaches with and without good surf breaks, *and* where the surf breaks aren't also near beaches that are good for lots of other recreational activities. People may pay premiums to live near beaches that happen to have nice surf breaks, but also are great for swimming, sunbathing, nature watching, or jogging. Differentiating the reasons why people live near a beach is extremely complicated when the beach is used for multiple types of recreational activities. In the Santa Cruz case study, the surf break that is examined is used almost exclusively for surfing and little else, due to the small amount of sandy beach adjacent to the cliffs.

But again, the site sample requirements to measure the contribution of proximity to a surf break to real estate values is challenging. It is only possible to estimate the added premiums of living near a surf spot within a coastal city, while the added premium of simply living in such a city to begin with is much more difficult to determine.

For example, in the Santa Cruz study, it is estimated that the premium in home values for living right next to a surf spot, relative to a mile away, is about $100,000; but this doesn't mean that someone living in the Santa Cruz mountains doesn't also pay a premium for being in relatively close proximity to the county's premier surf locations. The thought experiment is as follows: take the current Santa Cruz real estate market and then imagine what it would be if all of a sudden all of the surf breaks disappeared; no doubt the homes right near the most famous breaks would experience the greatest declines in value, but likely the drop in home values would sweep over the entire county, albeit at a rate inversely proportional to the distance from the coast.

Although it is not possible to generate precise estimates of what these city or county-wide real estate premiums are for being close to surf breaks, the Santa Cruz study suggests that they are significant. With almost 94,000 households in the county, if the average home premium to live near high-quality surf breaks was only $10,000 this would total to a county-wide real estate premium of $940 million. When multiplied by an average property tax rate of 1.1 per cent this results in county revenue of over $10 million per year. This revenue is sustainable indefinitely, as long as the surf persists.

What the Santa Cruz study suggests is that the consumer surplus (or much of it) of living close to surf breaks *is capitalized* into real estate values. This means that those fortunate enough to live close to surf resources end up paying for this privilege not through long commutes, but higher home prices (and by extension rental prices). Whether the higher real estate prices subsume the entire consumer surplus is unknown; answering this question would require very precise results from both TCM studies and HPM studies, which would be extremely difficult to obtain.

What are the research and policy implications?

To begin, since higher home prices significantly diminish any consumer surplus provided by proximity to surf breaks, and also have compounding effects on property tax revenues, we believe that this should be the direction that the non-market research of surfing takes. While examining the consumer surplus of surf spots through TCM is relatively straightforward, much more study is needed on the impact of real estate values through HPM. If researchers can conduct studies similar to the one in Santa Cruz, we will be able to get a much better sense of how much surfing impacts home prices (and rental markets), and begin to understand the much wider market effects surfing brings to local economies.

Examining the impact of surfing on property values is the first step in determining how much property tax revenue surfing generates for local communities. In regions where surfing is ubiquitous – California, Australia, Bali, South Africa – it is likely that the property tax revenues generated (indirectly) by surfing are large and growing, as surfing's popularity increases. However, in areas where surfing is relatively obscure or undeveloped, both the real estate price impacts and subsequent property tax impacts are likely negligible. But they are unlikely to remain that way.

Undeveloped surf locations, especially in remote locations like West Africa, will eventually be discovered and exploited, and the property values will appreciate significantly. Most of the money from the development of these areas will likely go to private land owners and developers. One way for local communities and governments to ensure that they receive a share of the wealth generated by these surf resources is to increase property taxes for properties proximate to surf breaks. Since access to these breaks is an extremely valuable public resource, for which congestion effects are almost guaranteed, there is a strong case to be made that property owners should pay for this privilege.[4] Raising property taxes is a simple and efficient way to channel money derived from privileged access to a unique public resource to the wider community.

Higher property taxes can address one piece of the management challenge faced in developing (and developed) countries that contain premier surf resources, but the larger issues of public access, sustainability, and local employment require specific planning and zoning requirements, as well as environmental regulations. What property taxes ensure is that even if the real estate proximate to surf resources is bought and owned by a small group of wealthy individuals (whether domestic or foreign), some of the value of the resource will continue to flow to the local community in perpetuity. Many countries have extremely low property tax rates, and some – including Namibia and Palau that have renowned waves – don't collect property tax at all; therefore, there is the potential to raise significant revenue using this policy mechanism.

In summary, research on the non-market value of surfing is relatively sparse, with no documented studies on the non-use value, and few studies on consumer surplus. To date, only one study has been published that examines the impact of proximity to surf breaks on home prices, which the authors argue likely subsumes a good deal of whatever consumer surplus exists for those surf resources. Given that there is likely a strong connection between proximity to surfing and real estate values, and the impact this has on property taxes, we believe that this area of research should be a priority for those interested in surfonomics.

Notes

1 Direct and consumptive values are values derived from use of a resource; even simply viewing surfing as a spectator is a form of consumptive use. Indirect and non-consumptive are values obtained without direct interaction with the resource.
2 The authors included two winners of the Nobel Prize in Economics and other leaders in the economics field.
3 In these contexts if there are few locals who use the surf breaks, then there aren't many people who potentially receive large consumer surplus from the resource; in addition, if these locals are low-income, their WTP is constrained, further reducing their consumer surplus. This is why TCM studies for surfing are typically conducted in areas where large majority native populations use the resource.
4 There is a more general case to be made that property values should be higher for all coastal property, but the case is particularly strong for unique resources like surf breaks, coral reefs, or MPAs.

References

Carson, R., R. Mitchell, M. Hanemann, R. Kopp, S. Presser, and P. Ruud (2003). Contingent valuation and lost passive use: damages from the Exxon Valdez oil spill. *Environmental and Resource Economics* 25, 257–286.

Chapman, D. J. and W. M. Hanneman (2001). Environmental damages in court: the American trader case. In A. Heyes (ed.), *The Law and Economics of the Environment*. Cheltenham, UK: Edward Elgar. pp. 319–367.

Coffman, M. and K. Burnett (2009). The value of a wave: an analysis of the Mavericks region – Half Moon Bay, California. Davenport, CA: Save the Waves Coalition.

Haab, T. C. and K. E. McConnell (2002). *Valuing Environmental and Natural Resources: The Econometrics of Non-Market Valuation.* Northampton, MA: Edward Elgar.

Lazarow, N. (2009). Using observed market expenditure to estimate the value of recreational surfing to the Gold Coast, Australia. *Journal of Coastal Research* 56, 1130–1134.

Murphy, M. and M. Bernal (2008). The impact of surfing on the local economy of Mundaka, Spain. Davenport, CA: Save The Waves Coalition.

Nelsen, C. (2012). *Collecting and Using Economic Information to Guide the Management of Coastal Recreational Resources in California.* Thesis. University of California, Los Angeles.

Nelsen, C., L. Pendleton and R. Vaughn (2007). A socioeconomic study of surfers at Trestles Beach. *Shore & Beach* 75(4), 32–37.

Pendleton, L. and J. Kildow (2006). The non-market value of beach recreation in California. *Shore & Beach* 74(2), 34–37.

Reed, D. (2011). Triple Crown generates 20.9M for HI. Editorial. *ESPN.* ESPN, 30 May 2011. Web. 8 November 2015.

Scorse, J., F. Reynolds and A. Sackett (2015). The impact of surf breaks on home prices in Santa Cruz, CA. *Tourism Economics* 21(2), 409–418.

Tilley, C. (2001). A valuation of the Pleasure Point surf zone in Santa Cruz, CA using travel cost modeling. California State University, Monterey Bay. 2007.

Part V

Reconceptualising sustainable surf spaces

9 Sustaining the local

Localism and sustainability

Lindsay E. Usher

Introduction

Sustainability is a term fraught with multiple meanings. Therefore, before discussing localism in the context of sustainability, sustainable surfing must be defined. Sustainability in surfing, much like the sustainability of other industries and processes, can be contradictory, depending on what aspect of sustainability to which one is referring. For example, the economic sustainability of surfing (i.e. perpetual income from the sale of surfboards and surfing gear, or perpetual income received by surf tourism communities) can be in direct conflict with the sustainability of surfing culture (i.e. maintenance of the traditions, values and social relationships of surfers). More people surfing creates crowding, which exacerbates conflict and threatens values which surfers try to foster, such as respect for one another or surfing elders (Booth, 2013; Daskalos, 2007; Kaffine, 2009; Waitt and Warren, 2008). Economic sustainability can also conflict with the environmental sustainability of surfing (i.e. the maintenance of quality surf breaks in clean water so they can always be surfed). Increasing numbers of people in a surf tourism destination signify the need for more infrastructure, such as the construction of hotels, more boats carrying surfers which can lead to more waste and pollution, and coastal developments which may physically alter surf breaks, deteriorating the resources surfers come to use (Hall, 2001; Pendleton, 2002; Ponting and O'Brien, 2015).

Another aspect of surfing sustainability which conflicts with economic sustainability is sustainable wave access (i.e. the continued availability of 'riderless' waves for surfers or their ability to physically access the surf break). Crowded line-ups often mean surfers catch fewer waves (sometimes no waves at all), or are unable to complete their rides because of the presence of other people. Tourism development of coastal areas can threaten surfers' ability to access the surf break: property owners will sometimes privatize beach accesses, which limits the number of locals' beach access points (Iatarola, 2013). As the previous example shows, economic sustainability is not the only type of sustainability which conflicts with other types. Therefore, sustainable surfing will not be discussed as one concept, but as the four concepts highlighted above: economic, cultural, and environmental sustainability, as well as sustainable wave access.

Localism

Surf localism is the territorial behaviour of local resident surfers over a nearby surf break (Stranger, 2011; Usher and Kerstetter, 2015a; Warshaw, 2003). Localism can include a wide range of behaviours and is considered to be on a spectrum from mild to heavy (Nazer, 2004). At the most extreme (heavy localism), local surfers will not allow visiting surfers to surf at a location, enforcing this through bodily harm and property damage (Kaffine, 2009; Nazer, 2004). At the other end of the spectrum – mild localism (which is much more common) – local surfers expect respect from visitors and may violate surf etiquette or give visitors dirty looks (Beaumont and Brown, 2014; Nazer, 2004; Preston-Whyte, 2002). In between, there is moderate localism, when visitors might be made to feel unwelcome through occasional verbal threats and surf etiquette violations (Nazer, 2004). Cash localism is another form of localism, when surfers have to pay to gain entry to a surf break (Alessi, 2009; Nazer, 2004). The Ranch, in California, is the best example of this: surfers must own land in order to surf there (Alessi, 2009; Nazer, 2004). Surf resorts in the Indo-Pacific have also used 'pay-to-play' policies but some of those regulations are collapsing under economic liberalization (Alessi, 2009; Ponting, 2014; Ponting and O'Brien, 2014).

It is important to distinguish between 'surf rage' and localism, terms which are often used interchangeably. Surf rage is the anger surfers feel over not being able to catch waves, due to crowds and people not following surf etiquette, and the ensuing violence (Stranger, 2011; Young, 2000). Part of what drives surf rage is the myth of the perfect wave promoted in surf media: empty surf line-ups where surfers do not have to compete with anyone for waves (Ford and Brown, 2006; Preston-Whyte, 2002; Scheibel, 1995). However, localism has an added dimension of hierarchy, which distinguishes it from basic aggression over resources (Nazer, 2004). Localism also implies a sense of ownership and attachment to a particular place and the need to defend it from outsiders (Daskalos, 2007; Usher and Kerstetter, 2015a). Surf rage can occur between two local surfers or two visiting surfers; localism denotes an insider versus outsider dichotomy. While surf rage can be the product of localism (Scott, 2003), the two terms should not be equated with one another. The difference between an angry surfer and an angry local surfer is the attachment the local feels for the place where they live and surf.

Surfers' sense of ownership and attachment to the places where they surf is incredibly strong (Lawson-Remer and Valderrama, 2010). Anderson (2014) explored the ways in which the places where people surf become part of their identity. Surfers form their identity through regular interaction with the ocean where they have grown up, learned to surf and have spent many hours surfing (Anderson, 2014). Surfers who always surf one particular break know when, where and how the waves will break there because they have surfed there so often (Comley and Thoman, 2011; Daskalos, 2007; Usher and Kerstetter, 2015a). This surfing knowledge is part of what makes them a 'local' of a surf break. The other part of being a local is the connection to other surfers and membership in the local surf culture (Beaumont and Brown, 2014; Langseth, 2012). Daskalos (2007), Dorset (2009) and Waitt and Warren (2008) all discuss how surfers' attachment to the surf break is not simply connected to the time they have spent there, but the bond they have formed with other surfers in that space.

Sustaining the local

Localism has always been about 'sustaining the local': maintaining the availability of waves for locals in a place to which they are strongly attached (sustainable wave access for locals) and maintaining the surf culture they have formed with other local surfers (sustainable local surf culture) (Scheibel, 1995; Sweeney, 2005). When crowds or outsiders threaten local wave access and surf culture, the need to defend the surfing space arises (Lawson-Remer and Valderrama, 2010; Scheibel, 1995). While some suggest localism originated in Southern California, when people from the San Fernando Valley came to the beaches to surf (this was when the term first appeared), Hawaiians were actually the first surfers who needed to defend their local surf space and culture (Ingersoll, 2009; Scheibel, 1995; Stranger, 2011; Walker, 2011).

Surfing originated in Hawai'i and scholars have pointed out that Hawaiians demarcated surfing territory: certain breaks were for royalty and other people had to go elsewhere (Alessi, 2009; Sweeney, 2005). Ingersoll (2009) discussed the Hawaiian *kapu* system, which was similar to localism in that it established a pecking order and etiquette in the water surfers had to follow. When Hawai'i was colonized, surfing was discouraged by missionaries and the numbers of surfers dwindled but many people still surfed (Walker, 2011). When American colonists tried to claim surfing as their own and began earning profits through the promotion of surfing for tourism at the beginning of the twentieth century, Hawaiians resisted by starting their own surfing businesses, which were successful and helped them maintain control over their surf culture and space (Walker, 2011).

In mid-1970s, there was an incident between Australian surfers, including Rabbit Bartholomew, and Native Hawaiians. The Australians had bragged to the surf media of their dominance over Hawaiian waves, contests and surfers and Native Hawaiians felt disrespected by these claims on their surfing space. This tension resulted in a violent confrontation with Bartholomew and ended with Hawaiian Eddie Aikau bringing the Australians and Hawaiians together in a make-shift court in order to resolve their differences (Bartholomew and Baker, 2002; Walker, 2011). These ways of resisting claims on their surfing space and culture were ways in which Hawaiians could still exert control over ocean territory as they lost control over land territory to tourism and surf industry development (Walker, 2011). Similar to Walker, Ingersoll (2009) agreed that the modern Hawaiian locals imposed localism in order to recapture indigenous surfing space taken from them by the neo-colonial surf tourism industry. While colonization and its consequences were the greatest threats to Hawaiian surf culture and space, localism in other places has been blamed on two other connected factors: commercialization and crowding (Booth, 2013).

Threats to the local: commercialization and crowding

The commercialization of surfing began in the 1960s when surf movies and music reached mainstream audiences and surfboards started becoming mass produced (Booth, 2013; Ormrod, 2005; Scott, 2003). The boards were smaller and made of lighter material, thus making surfing accessible to greater numbers of people (Daskalos, 2007; Ormrod, 2005; Scott, 2003). Everyone wanted to try surfing, even if they did not live near an ocean (Ormrod, 2005). Over the decades, competitive surfing became professional surfing, which relied on the profits of surf companies which had grown into large corporations through the promotion of the sport to the masses (Booth, 2013). Lanagan (2003) dubbed it 'surfing capital': at the centre of which are the major surfwear companies (Rip Curl, Billabong and Quiksilver), but also includes peripheral apparel companies which commodify surfing. Surfing became an economically sustainable industry to be a part of; it was no longer a fringe, transgressive activity (Ormrod, 2005). It remains a popular sport: surf movies continue to enter the mainstream, the surf clothing industry is still big business and surf schools are thriving in many coastal communities (Lanagan, 2003; Stranger, 2011). While the commercialization of surfing has made surfing economically sustainable, increasing crowds for surfers in the United States, Australia and other countries over the last half century have threatened the sustainability of local surf culture (Booth, 2013; Ormrod, 2005; Stranger, 2011).

Many researchers believe crowding is at the root of localism (Booth, 2013; Daskalos, 2007; Scott, 2003; Stranger, 2011). Local surfers are threatened by crowding because it means they have to share 'their' wave resource with many other people, some of whom might not know surfing rules or have very much experience (Daskalos, 2007; Stranger, 2011). However, crowding is not just a threat to the sustainability of local surf culture and wave access, it threatens the wave access of all surfers (because it reduces the number of waves each surfer catches and often, the length of their rides) and can lead to the aforementioned problem of surf rage (Booth, 2013; Stranger, 2011).

Multiple researchers have used the concept of common pool resources to examine surfing (Alessi, 2009; Nazer, 2004; Ponting and O'Brien, 2015; Rider, 1998). While surfers are competing over an inexhaustible resource, there are limited places in the world where the right set of conditions (weather, swell, geography, bathometry, tides etc.) come together to generate quality waves for surfing (Alessi, 2009; Nazer, 2004). Therefore not every place with a beach has waves; surfers are limited in where they can surf. Increasingly crowded line-ups in these limited places with quality surf have resulted in dangerous conditions for everyone. Surfers have to manoeuvre around other surfers to avoid hitting them, surfers can become violent if they are angry over not catching waves due to the number of people in the water, and a large number of people with large objects attached to them in decent surfing conditions can quickly become unsafe (Stranger, 2011; Young, 2000).

One suggestion for alleviating crowding has been the construction of artificial surfing reefs (creating waves where there were none naturally) but it is controversial (not necessarily environmentally sustainable) and may not solve the need for more quality surfing space or more days during the year in which surfers can surf (Alessi, 2009; Rendle and Rodwell, 2014; Stranger, 2011). An assessment of the Boscombe Reef in England, where wind and swell do not generate prime surfing conditions to begin with, found that providing space for less-than-ideal waves to break did not necessarily increase the number of surfable days, therefore surfers quickly lost interest in it (Rendle and Rodwell, 2014). Surf travel is another way in which surfers have tried to avoid crowds, but even many remote locations are now becoming crowded as they become more well-known (Ponting and O'Brien, 2015). The other way in which surfers cope with crowds is an unwritten system of rules known as surf etiquette (Alessi, 2009; Lawson-Remer and Valderrama, 2010; Nazer, 2004).

Surf etiquette is a system of rules surfers have developed in order to try to share the wave resource with one another and keep everyone safe (Alessi, 2009; Ford and Brown, 2006; Lawson-Remer and Valderrama, 2010; Nazer, 2004). Surf etiquette includes: not 'dropping in' on the surfer with the right-of-way (person furthest on the inside of the wave – closest to the breaking part of the wave), taking turns for waves, not paddling out in the path of oncoming surfers and maintaining control of one's board (Nazer, 2004). Ideally, if everyone follows the rules, there is sustainable wave access: everyone gets enough waves and there is no conflict (Ponting and O'Brien, 2015). Multiple researchers advocate for this ethical, virtuous system of reciprocity (Alessi, 2009; Olivier, 2012; Sweeney, 2005). After a well-publicized fight between two Australian surfers, Nat Young and another local, some surfers came together to try to formalize the unwritten rules of surfing and place them on placards at popular surf breaks (Stranger, 2011). This movement has spread to the United States and Europe as well. Surf etiquette brings some order to the chaos of crowded surf breaks. However, localism changes the rules of the game.

Many locals believe that an additional rule of surfing is that they have priority, no matter who caught the wave first, or who is further inside (Lawson-Remer and Valderrama, 2010; Rider, 1998). Therefore, they can take off on a wave when they want. Local surfers will even violate surf etiquette in order to assert dominance or because they think it is their right (Lawson-Remer and Valderrama, 2010; Preston-Whyte, 2002; Usher and Kerstetter, 2015a). Another rule is that visiting surfers must show deference and respect to locals: they must work their way into the line-up, they should not go straight to the peak and expect to catch the biggest waves (Lawson-Remer and Valderrama, 2010; Nazer, 2004).

While these 'local rules' appear to complicate surf etiquette and possibly create unsafe situations, another aspect of having a local surf crew is that they will help enforce these rules and other aspects of surf etiquette (Lawson-Remer and Valderrama, 2010; Stranger, 2011). Locals may ask surfers that violate the rules to leave the surf break (Nazer, 2004; Stranger, 2011). Thus, the locals are the ones in charge of keeping everyone safe by making sure everyone follows the rules. This local regulation leads to the sustainability of local surf culture, because the locals are in control of the surf break and visitors follow the rules of the culture they have created, and also ensures sustainable wave access for locals, because of their assumed priority in the line-up. Provided visiting surfers show respect for, and defer to, locals, locals are more likely to respect them and give them more waves, thereby increasing their wave access as well (Beaumont and Brown, 2014; Dorset, 2009; Usher and Kerstetter, 2015a). While local regulation has both negative and positive implications, the following section discusses other positive implications of localism for the cultural and environmental sustainability of surfing.

Sustaining the local: culture and environment

As previously mentioned, one of the original purposes of localism was cultural preservation (or sustaining the local surf culture, including the order locals had established in the waves) (Olivier, 2012; Scheibel, 1995; Sweeney, 2005). Hawaiians' resistance to the foreign appropriation of surfing and surf culture in Hawai'i, beginning in the early 1900s, is one example of this (Walker, 2011). In later efforts to reclaim their surfing space in the 1970s, Hui O He'e Nalu, a club representing Hawaiian surfers, negotiated for Hui members to have positions as water patrolmen in order to earn income from the International Professional Surfing contests which were held on the North Shore (Walker, 2011). The club used the income to benefit the local community, which helped preserve the local community and culture. They created surfing contests for local surfers, gave away surfboards as prizes and organized community festivals for local Hawaiian children and families (Walker, 2011). Hawaiians' organized resistance to the surf industry increased the sustainability of their own surf culture and community. Localism can also be a source of cultural empowerment for locals who could be part of disenfranchised communities on land (Usher and Kerstetter, 2015a; Walker, 2011). Locals, such as surfers in Hawai'i and Bali, are able to feel ownership for the waves and maintain control of actions in the surf even if they cannot control powerful development forces on land, such as inflated costs of basic goods and the displacement of their communities due to tourism development (Leonard, 2007; Usher and Kerstetter, 2015a; Walker, 2011).

Surfers also will try to preserve local surf culture by preventing outsiders from accessing the surf. In the 1960s in California, one group of local surfers around Los Angeles attempted to block the roads to the coast from the San Fernando Valley so that 'Vals' (surfers from the valley) could not access the beach (Scheibel, 1995). Surfers wanted to protect the surf breaks from 'Vals', who would contaminate the local surf culture, alter the order locals had established in the water, and threaten locals' wave access (Scheibel, 1995). By naming these outsiders, they further asserted dominance by controlling the language (Scheibel, 1995). However, in some cases, naming works to bring outsiders into the local surf culture. When resident foreigners assisted local residents in Nicaragua, such as offering low-cost board repair, or helping families in need, local surfers considered them part of the local surf community (Usher and Kerstetter, 2015a). Unlike California, not all outsiders were thought to threaten the local way of life; some contributed positively towards sustaining the community and local surf culture, therefore calling them 'local surfers' acknowledged that contribution (Usher and Kerstetter, 2015b). These outsiders strengthened the local surf culture through their positive contributions and also strengthen it through increasing the number of members of the local culture.

Localism can also make positive contributions to the environmental sustainability of surfing. Surfers are some of the most fervent environmental activists due to their regular interaction with the ocean (Wheaton, 2007). The high level of ownership local surfers feel for their home break also encourages environmental conservation (Lawson-Remer and Valderrama, 2010; Sweeney, 2005). Lawson-Remer and Valderrama (2010) discussed ways local surfers have mobilized in order to protect their local waves from environmental threats such as marine and beach developments which can alter the formation and consistency of surf breaks. The strong bonds surfers form surfing with one another on a regular basis also help to facilitate these organized movements (Lawson-Remer and Valderrama, 2010). Surfers have mobilized communities in California, Puerto Rico and England to prevent the destruction of world-class surf breaks and advocate for clean beaches and oceans (Lawson-Remer and Valderrama, 2010; Pendleton, 2002; Wheaton, 2007).

Localism originated in the need to protect local surf culture and wave access. Colonialism, commercialization and crowding have been the biggest threats to sustaining local culture and access. Locally enforced surf etiquette can be a positive aspect of localism which helps to sustain wave access for locals and visitors. Localism also fosters local surf culture preservation and empowerment in multiple ways, as well as environmental protection. While these examples highlight how positive localism can be, localism is largely perceived as a negative aspect of surfing. The following discussion explores the ways in which localism challenges, and complicates, aspects of sustainable surfing.

The challenges of sustaining the local

Surfers deal with a wide variety of preconceived notions people have about them: lazy, marijuana-using, laid back, beach bums, and countercultural societal drop-outs are just some of the stereotypes associated with surfers (Caprara, 2008; Kaffine, 2009; Pearson, 1982). Popular press coverage of surf rage incidents and YouTube videos of locals beating up other surfers portray an image of surfers as violent assailants as well (McHugh, 2003; Young, 2000). While surfers may appreciate this because it discourages more people from joining the line-up, negative public perceptions of surfers and surfing obscure the benefits of surfing and the positive contributions surfers have made to their communities. Organizations such as Surfing for Autism and partnerships between the United

States Surfing Federation and Wounded Warriors have provided positive experiences for people with disabilities (Surfing for Autism, 2015; United States Surfing Federation, n.d.). However, instead of seeing surfers as role models, people may see the negative images of surfers and be discouraged from participating in the sport.

Despite negative publicity, however, surfing remains a popular activity. Surf schools are currently proliferating worldwide, including many geared towards women specifically (Stranger, 2011). In the past, there were no surf schools: learning to surf was not easy and beginners had to find equipment and someone to teach them (Stranger, 2011). Before surfing was such a trendy sport, local surfers would assist newcomers in learning a surf break (Booth, 2013). However, now, the large number of people who have learned to surf, and are crowding surf breaks, has discouraged surfers from helping out newcomers (Booth, 2013). If they help them, they become another person with whom the surfer will have to compete for waves. While there are plenty of surf schools where beginners can learn, achieving a higher level of surfing skill takes spending a considerable amount of time, patience and effort in the water (Mendez-Villanueva and Bishop, 2005; Nourbakhsh, 2008; Sotomayor and Barbieri, 2015). As one of Nourbakhsh's (2008) participants described: 'It seems like surfing is a sport that never ends ... there is always something to work on. And when you think you got it ... you go the next time and it's like you are a beginner again' (2008: 91).

Outside of the protected environment of a surf school, many of which operate in areas with smaller, gentler waves, beginners encounter crowds, aggression and surf conditions for which they may be unprepared (Stranger, 2011). Beginners must learn surf etiquette, how to read waves and how to navigate the board once they can stand (Langseth, 2012). Many surfers learn surf etiquette from other surfers who inform them of their mistake after they have violated the rules: it is part of the socialization process which surfers go through (Langseth, 2012; Usher and Kerstetter, 2015a). If beginners arrive at a surf break with unfriendly locals who do not care to assist them, or are rude to them, they could quickly become discouraged from continuing the sport (Langseth, 2012). They could also be a danger to themselves and others if experienced locals are not willing to inform them of surf etiquette or information about the surf break. Another discouraging aspect for beginners is the idea of having to work their way up in the local surfer hierarchy (Nazer, 2004). Beginners must spend much time in the water to improve their surfing and earn the respect of their fellow surfers, many of whom may have been surfing for many years. Therefore, while surf shops may profit off the high number of casual surfers (interested beach-goers who explore surfing as one of many activities) they bring in through surf lessons, casual surfers are less likely to become more dedicated recreational or hard-core surfers, who return to buy more advanced (expensive) surfboards and gear, if they undergo the aforementioned discouraging experiences (Orams and Towner, 2013).

However, one argument for this long learning process is that it may actually cull the herd and weed out the people who are not dedicated to the sport. This would make for less crowded conditions, fewer conflicts in the water and greater safety for everyone. The ones who are dedicated may earn respect from locals through the amount of time they dedicate to learning to surf and improving (Langseth, 2012; Nazer, 2004). If these dedicated beginners learn the rules and respect the locals, locals will be more likely to help them to become competent surfers and they would be more motivated to continue surfing. Orams and Towner (2013) refer to these surfers as recreational surf-riders (or surfers). While surfing is not the primary focus of their life, they are passionate about it, and their ability levels range from competent to expert, and they surf a variety of different surf breaks (Orams and Towner, 2013). Recreational surfers will keep coming back to surf shops to swap and buy boards and more advanced gear. They are the repeat customers whom surf shops desire because they have to buy boards and gear, unlike hard-core surfers who might receive free boards and gear because they are professional surfers or sponsored (Orams and Towner, 2013). Therefore, the economic sustainability of surfing would depend upon supportive local surfers with years of experience willing to guide dedicated beginners. Beaumont and Brown (2015) found that surfing was a family activity: parents encouraged their children to surf and everyone would go to surf contests together. Social connections greatly influenced people's participation in surfing (Beaumont and Brown, 2015). If surfers are part of a community, or find a community, that supports them and helps them grow and improve, they will likely continue in the sport: these bonds with other surfers are part of what tie them to a surf break (Beaumont and Brown, 2014; Waitt and Warren, 2008).

Another aspect of the economic sustainability of surfing is the viability of surf tourism destinations. Being able to surf alone and avoid the increasing crowds at surfers' home breaks has been a major impetus for surf tourism, which has become a booming industry (Alessi, 2009; Ormrod, 2005; Ponting, 2009). *The Endless Summer* opened up the world of international surf travel to American surfers in 1964 (Ormrod, 2005). There were empty waves to be conquered and surfed in unknown places. When surfers travelled, they interacted with local residents, who began to surf as well (Usher and Kerstetter, 2015a). One could probably find a local surfer in nearly every county with waves now: the International Surfing Association has 96 member countries (International Surfing Association, 2015). Localism becomes complicated in a surf tourism context, especially in developing countries.

In many places, local residents are highly dependent upon the surf tourism industry which develops: tourism starts to compete with agriculture and marine fisheries as the primary economic generator (Usher and Kerstetter, 2015a). If heavy localism develops, local surfers could lose the income upon which they have grown increasingly dependent as tourism envelopes more of the local economy (Usher and Kerstetter, 2015a). Surf tourists do not want to be harassed where they travel, so they may seek out other destinations (Usher and Kerstetter, 2015a). Places such as Hawai'i, El Salvador and parts of Costa Rica are known for unfriendly locals. While Hawai'i will always be a popular surf location because of its reputation as the birthplace of surfing, surfers have started going further out into more remote areas of the world to avoid crowds and spots with heavy localism. In some places, such as the United States and Australia, many local surfers are not necessarily dependent upon tourism. They work in other industries so they can afford to repel visitors. Some expatriates who relocate to developing countries do not have to work in tourism either: if they have money saved up, go back to their country to work for part of the year, or are independently wealthy, they might not have to work at all since their money goes further in the developing world. Unlike local residents, most of whom are at a greater economic disadvantage, these expatriates can afford to enact localism against visiting tourists. This could endanger the economic sustainability of the surf tourism destination, if the localism becomes extreme enough.

Localism has a complex relationship with the sustainability of surfing. It can create a negative image for the sport of surfing and discourage beginners from pursuing the sport, threatening the economic sustainability of the sport. However, localism may encourage dedicated beginners to continue surfing, which would support the economic sustainability of surfing. Economic sustainability is further complicated by the proliferation of surf destinations. Localism may repel tourists from some destinations. The following section discusses research on localism, which provides additional support for the complex relationship between localism and sustainability.

Sustaining the local: observations from the field

Research on localism shows the myriad ways in which it manifests in different places as local surfers try to protect their local surf culture and wave access. For example, Nazer (2004) noted that heavy localism does not always develop in crowded urban beaches: Oregon and Washington have some heavily localized beaches. It does not seem to matter what the socio-economic income level is either: localism has been observed in working-class areas, as well as wealthy beachfront communities (Nazer, 2004). Kaffine (2009), in the largest scale study of localism using data from the ratings of different factors at surf breaks on the Surfline website for the state of California, found that the level of localism was highly correlated with the level of wave quality at a surf break. Localism was also highly correlated with the amount of congestion (crowding) at a surf break. Level of paddling difficulty was negatively correlated with localism, suggesting that more advanced surf breaks attract more experienced surfers and there is less need for regulation (Kaffine, 2009). In a study of surfers in Southern California, Comely and Thoman (2011: 24) found that 'surfers' perceptions of territoriality and waves as limited resources successfully predicted surf-related negative affect, even when controlling for individual differences in general aggression tendencies'. Surfers' close connection with the space predicted surf-related aggression.

Anecdotal evidence in much of the localism literature cites violent incidences between surfers in Australia: one of the most notorious surf gangs, the Bra Boys, are from Maroubra, which is a suburb of Sydney (Nazer, 2004; Stranger, 2011;

Young, 2000). In Australia, Waitt and Frazer (2012) found localism to be connected with masculinity and forming bonds with other surfers, much like Waitt's previous work (Waitt, 2008; Waitt and Warren, 2008). However, they also observed differences between the ways in which shortboarders and longboarders would territorialize the surf break. Longboarders complained about the ways in which shortboarders, typically younger surfers, would try to exclude them from the surfing space. Although, at least one longboarder also spoke about defending his surfing space from newcomers, who were immediately obvious once they paddled out (Waitt and Frazer, 2012).

Researchers in New Zealand and England discovered a milder form of localism (Beaumont and Brown, 2014; Dorset, 2009). Local surfers in one part of England felt a deep bond between their local community and the surf break; they expected visitors to respect them and felt that they could violate surf etiquette due to their local status (Beaumont and Brown, 2014). This benign localism did not create many problems; however, crowding exacerbated tensions and the need to violate surf etiquette. Researchers also noted that along the same coastline, there was a 'locals only' beach where heavier localism was present (Beaumont and Brown, 2015). Dorset (2009) had similar findings from several beaches in New Zealand: local surfers would violate surf etiquette to show dominance. In several beaches, locals did not mind sharing waves but in other beaches, they would share on the condition that visitors respected the locals and followed the rules. Not following the rules resulted in verbal abuse (Dorset, 2009). Both of these studies indicate the importance of placing localism on a spectrum as Nazer (2004) suggests, instead of defining it strictly as heavily aggressive and violent behaviour on the part of locals.

Much of the previous work focuses on localism in the developed world. However, there have been observations of localism in the developing world, where many surfers have travelled to explore new surf breaks and avoid the crowds. Localism takes on a new meaning for surfers in the developing world, because many of them are reliant upon income from tourists to support themselves and their families (Usher and Kerstetter, 2015a). There is also speculation that surfers learned localism from the Western surfers who introduced the sport and taught them how to surf, it perhaps may not have developed on its own (Young, 2000). Local surfers in Kuta, Bali were introduced to surfing by Western tourists, but they now control the surf lineup and decide who gets to take which waves. One surfer seemed to suggest they had previously been dominated through colonialism but now they were the ones who controlled the surfing space (Leonard, 2007). This is similar to the empowerment Hawaiians have gained from resistance to colonialism through localism (Ingersoll, 2009; Walker, 2011). Other studies have found that local surfers in developing countries are not the ones enforcing localism, they want to share their wave resource with visitors; resident foreigners are the ones who enact heavy localism instead (Krause, 2013; Usher and Kerstetter, 2015a). This brings an added dimension to localism: the length of time someone has to live or surf somewhere to be considered a local.

Much of the discussion of localism takes who 'local surfers' are for granted: they simply live near the wave under conflict. Comley and Thoman (2011: 16) define a local as someone who 'has lived and surfed in the area for a long period of time'. Scott (2003) notes another definition which has been put forward in the *Surfin'ary* which is, 'anyone who's been there a day longer than you'. Langseth (2012) discussed the conditions under which Norwegian surfers would start to consider new residents locals, and earning respect through the demonstration of surfing skill is part of that. Ishiwata (2002) discussed how Hawaiian surfers used to be able to distinguish locals based on their boards, but since many Hawaiian brands have gone global, it was difficult to do so. 'Real' locals used boards from underground, obscure shapers and that was how they could distinguish one another (Ishiwata, 2002). As previously mentioned, many foreigners live in developing countries in order to be near high-quality surf conditions and become resident surfers in those places. We (Usher and Kerstetter, 2015b) explored this complex identity formation in Nicaragua. While some foreigners considered themselves to be local surfers, the indigenous local surfers did not necessarily consider them local surfers (Usher and Kerstetter, 2015b). This added complexity complicates localism even further: if there are multiple groups of locals, they may differ in how they treat visiting surfers and regulate the surf break. There may be multiple local surf cultures present. This can create conflict between residents, as well as between residents and visitors.

As the above research demonstrates, there are similarities and differences in the ways localism manifests in different places. Localism is present in many locations, and ranges from mild to heavy. While many surfers travel to escape their increasing crowded home breaks, localism affects surf destinations in complicated ways as well, especially in the developing world. The case studies presented below provide more detailed comparative evidence from two surf tourism destinations.

Localism in Nicaragua and Costa Rica

The following is based on research on localism I conducted in Nicaragua (Usher and Kerstetter, 2015a, 2015b) and Costa Rica (Usher and Calhoun, 2015; Usher and Gómez, 2016). Both of these countries are at different stages of surf tourism development. Costa Rica has long been known as a surf destination and Nicaragua is up and coming (Krause, 2013; Weisberg, 2010). While there are many similarities within surfing communities in the two countries, the differences could be attributed to the amount of time each place has been known as a surfing destination. The two surf destinations presented here are Popoyo in Nicaragua (a reef break on the southwest Pacific coast) and Pavones in Costa Rica (a point break on the Pacific coast famously known as the second longest left-breaking wave in the world).

I conducted ethnographic research on localism in both communities. I spent two months in Las Salinas, Nicaragua in May to July of 2012. I spent two months in Pavones, Costa Rica, from May to July of 2014. In both communities, I lived with a local family and integrated myself into the local surf culture.

I obtained the samples in both communities through snowball sampling: as I got to know people in the community, I would ask them to refer me to other surfers, local and foreign, to whom I could speak (Kuzel, 1999). In Nicaragua, I interviewed 23 local surfers, 14 resident foreigners and 27 surf tourists. In Costa Rica, I interviewed 24 current and former local surfers, 31 resident foreigners and 35 surf tourists. I speak Spanish, therefore I conducted the interviews in the language participants were most comfortable using. I also engaged in participant observation in both communities: I took field notes after surfing and kept a journal of my other experiences and observations throughout my time there. Data analysis occurs throughout the process of data collection in ethnographic research (LeCompte and Schensul, 1999). Therefore, as certain themes developed throughout my interviews and observations in both field sites, I looked for participants who could provide disconfirming viewpoints (Kuzel, 1999). The interviews were all transcribed into the participants' native language. The transcripts were analysed for themes of territoriality and identity but in both studies, researchers remained alert to additional themes that might also be present. Two other researchers reviewed my field notes or interview transcripts as part of the peer de-briefing process (Creswell and Miller, 2000).

Both destinations have two groups of resident surfers: locals and resident foreigners. As previously discussed, this is an added dimension which complicates localism in many developing countries. Nicaraguans had a variety of opinions about who they considered local and who they did not: resident foreigners who were friendly and assisted community members were considered local surfers by some of the indigenous surfers. Most Costa Ricans did not consider resident foreigners as local surfers, only Costa Ricans were locals. Resident foreigners in Nicaragua had diverse opinions on their local status: a few considered themselves to be local surfers but others said they were not because they had not been born there. The majority of resident foreigners in the Costa Rica study considered themselves to be local surfers, due to the time they had lived and surfed in Pavones. Many of them had lived in Costa Rica for ten to 20 years or longer. Most resident foreigners had not been in Nicaragua for more than 12 years.

Nicaraguan surfers expressed great feelings of ownership for the waves, but also the need to share it with others. Most Costa Ricans felt they had a right to the wave because they were native residents of the community. This strong sense of attachment was similar to surfers throughout the world (Anderson, 2014; Daskalos, 2007). In terms of regulating the surfing spaces, Nicaraguans said they would help visitors by reminding them of surf etiquette or explaining the surf break to them. They also said they had to tell visitors to calm down and not become aggressive. Costa Rican locals offered similar advice to tourists. Much like the mild localism which has been observed in other places, resident foreigners and tourists observed that Nicaraguans and Costa Ricans would blatantly violate surf etiquette by dropping in on people (Nazer, 2004; Preston-Whyte, 2002). Tourists and resident foreigners in both countries said that local surfers were generally friendly but dropping in was their major offence. However, resident foreigners in Costa Rica also complained that tourists were bad about dropping in on people, and that this forced them, and the locals, to drop in on tourists in return.

Ownership of the waves for resident foreigners in Nicaragua and Costa Rica manifested in a variety of ways. In Costa Rica, some used the term 'home break' instead of characterizing it as ownership. However, in terms of expressing ownership, some resident foreigners would get into verbal arguments with tourists and some had been in physical altercations in the past. Costa Ricans, other foreign residents and tourists all said that resident foreigners were much more aggressive and they had seen more problems with them in the water than with local Costa Ricans. This was similar to Nicaragua: Nicaraguans, resident foreigners and tourists described verbal altercations they had seen or experienced with resident foreigners. In both Popoyo and Pavones, surfers described physical fights that had occurred between particularly aggressive resident foreigners and local surfers. Some resident foreigners in both places seemed to believe they had the right to fight with anyone over waves. Other studies have also observed this heavier localism on the part of resident foreigners in developing countries (Krause, 2013). As previously mentioned, heavy localism on the part of resident foreigners has the potential to endanger local economic sustainability if it repels tourists.

Interestingly, in both places, surfers positively compared the surf destinations to other places in terms of the level of localism. Tourists in both places seemed to think there was a good vibe in the water and spoke highly of the locals. Nicaragua was positively compared with places such as El Salvador, which had a bad reputation for localism. At Popoyo, many locals said the locals at a beach a few miles away were much more aggressive and fought with tourists. Locals, resident foreigners and Costa Ricans kept emphasizing that Pavones was not like other places in Costa Rica, where locals were more aggressive. Tourists positively compared Pavones to places in the United States where localism was worse. With such positive ratings from tourists, it seemed that neither destination was in danger of becoming marked as a 'heavy localism' spot, which tourists would want to avoid.

Crowding was a bigger issue in Pavones than in Popoyo; however, they are very different waves. Ford and Brown (2006) note how the type of wave can affect the carrying capacity of the wave. Popoyo is a reef break that cannot hold as many people as the one kilometre long Pavones break. However, at Popoyo, two surfers can take the same wave: one can go right, the other can go left. Due to its smaller size, Popoyo feels more crowded with fewer numbers of surfers. Many people go to Nicaragua to avoid the crowds of Costa Rica but if they arrive there and find a crowd, they could look elsewhere for their next vacation. Local surfers have learned how to cope with crowding by surfing further on the inside of the break or other waves around the main break when there are too many people. They also know the wave better than the tourists, which allows them to be in the right spot and catch more waves even when there is a crowd.

While the Pavones break can hold more people, swell forecasts have drastically changed the way in which surf tourism functions in the town. Pavones can range from 25 people in the water one day to 80 the next if a large swell is predicted. Older locals said there were not fights over the waves in the early days because there were fewer people. However, something which some locals and resident foreigners noted was that localism had actually been worse in years past. Even though the crowds had increased, localism had actually decreased, which contradicts previous suggestions that localism is a result of crowding (Booth, 2013; Scott, 2003). Some resident foreigners complained of the way in which the more recent larger crowds of people exacerbated tensions in the water: everyone became more aggressive and there were more confrontations. This might suggest Pavones is actually dealing with surf rage instead of localism. The 'flash mob' surf tourism Pavones is experiencing based on swell forecasts could be incredibly detrimental if it is not controlled. While it is a famous wave, there are plenty of other quality waves which are much less crowded.

Popoyo and Pavones share a number of similarities in the way localism manifests in each surf destination. Local surfer identity is complex in each place because there are two groups of resident surfers. Local (native-born) surfers in both places seemed to be aware of the importance of sharing the waves to which they were strongly attached with tourists. Some resident foreigners in both places were more aggressive towards locals and tourists than local surfers. The resident foreigner localism could be a threat to tourism in both places, especially as more foreigners move to Nicaragua. While there was regulation of both surf breaks by locals and resident foreigners through advice to tourists and violations of surf etiquette, overall heavy localism did not appear to be a problem in either place. Tourists in both Popoyo and Pavones positively compared the localism in both destinations to other places, remarking on the good vibe they felt in the water. Crowding was a major issue at Pavones, due to the swell forecasts drawing in a large amount of tourists for a brief period of time. However, crowding has the potential to affect Popoyo in the future since its capacity is not as great. Crowding and the resultant conflicts appear to be greater threats to the economic sustainability of both destinations than localism.

Conclusion

Localism is about sustaining the local: preserving the local surf culture and protecting locals' wave access. Local surfers' culture and wave access has been threatened by colonization, commercialization and crowding. Surf etiquette is one of the primary ways in which local surfers are able to maintain control over their surfing space despite these challenges. While historically viewed as a negative phenomenon, localism facilitates both cultural and environmental sustainability in surfing, both of which are positive. However, localism can complicate economic sustainability in surfing, through perpetuating a negative image of surfing or discouraging beginner surfers. Research highlights the complexities and variations of localism in places where surfing has been popular for decades and surf destinations where it is growing in popularity. Two surf destinations, Popoyo, Nicaragua and Pavones, Costa Rica, are experiencing the mild localism present at many surf breaks; however, crowding has the potential to threaten the economic sustainability of both destinations in the future.

While localism may challenge the economic sustainability of surfing, sustaining the local surf culture and local surfing space is important for surf communities. Positive forms of localism should outshine this phenomenon which is known as the dark side of surfing. By encouraging positive manifestations of localism and discouraging the negative ones, surfers can work towards a more sustainable form of surfing.

References

Alessi, M. D. (2009). The customs and culture of surfing, and an opportunity for a new territorialism? *Reef Journal, 1*(1), 85–92.

Anderson, J. (2014). Surfing between the local and the global: Identifying spatial divisions in surfing practice. *Transactions of the Institute of British Geographers, 39,* 237–249.

Bartholomew, W. R. and Baker, T. (2002). *Bustin' down the door* (2nd edn). Sydney: HarperSports.

Beaumont, E. and Brown, D. (2014). 'It's not something I'm proud of but it's ... just how I feel': Local surfer perspectives of localism. *Leisure Studies,* 1–18. doi: http://dx.doi.org/10.1080/02614367.2014.962586.

Beaumont, E. and Brown, D. (2015). 'Once a local surfer, always a local surfer': Local surfing careers in a southwest English village. *Leisure Sciences, 37,* 68–86.

Booth, D. (2013). History, culture, surfing: Exploring historiographical relationships. *Journal of Sport History, 40*(1), 3–20.

Caprara, P. (2008). Surf's up: The implications of tort liability in the unregulated sport of surfing. *California Western Law Review, 44*(2), 557–587.

Comley, C. and Thoman, D. (2011). *Fall in line: How surfers' perceptions of localism, territoriality and waves as limited resources influence surf-related aggression.* Paper presented at the 91st Annual Convention of the Western Psychological Association, Los Angeles, CA.

Creswell, J. W. and Miller, D. L. (2000). Determining the validity in qualitative inquiry. *Theory Into Practice, 39*(3), 124–130.

Daskalos, C. T. (2007). Locals only! The impact of modernity on a local surfing context. *Sociological Perspectives, 50*(1), 155–173.

Dorset, W. (2009). *Surfing localism and attachment to beach space: A Christchurch, New Zealand, perspective.* (BA), University of Canterbury, Christchurch, NZ.

Ford, N. and Brown, D. (2006). *Surfing and social theory: Experience, embodiment and the narrative of the dream glide.* New York: Routledge.

Hall, M. C. (2001). Trends in ocean and coastal tourism: The end of the last frontier? *Ocean & Coastal Management, 44,* 601–618.

Iatarola, B. (2013). Surf tourism: Social spatiality in El Tunco and En Sunzal, El Salvador. *The International Journal of Sport and Society, 3,* 219–227.

Ingersoll, K. E. (2009). *Seascape epistemology: Decolonization within Hawai'i's neocolonial surf tourism industry.* (Doctor of Philosophy), University of Hawai'i, Honolulu, HI.

International Surfing Association. (2015). Member directory. Retrieved 9 September 2015, from www.isasurf.org/membership/member-directory/.

Ishiwata, E. (2002). Local motions: Surfing and the politics of wave sliding. *Cultural Values, 6*(3), 257–272.

Kaffine, D. T. (2009). Quality and the commons: The surf gangs of California. *The Journal of Law and Economics, 52*(4), 727–743.

Krause, S. (2013). Pilgrimage to the playas: Surf tourism in Costa Rica. *Anthropology in Action, 19*(3), 37–48.

Kuzel, A. J. (1999). Sampling in qualitative inquiry. In B. F. Crabtree and W. L. Miller (eds), *Doing qualitative research* (2nd edn) (pp. 33–45). Thousand Oaks, CA: Sage Publications.

Lanagan, D. (2003). Dropping in: Surfing, identity, community and commodity. In J. Skinner, K. Gilbert and A. Edwards (eds), *Some like it hot: The beach as a cultural dimension* (pp. 169–185). Aachen: Meyer & Meyer Sport.

Langseth, T. (2012). Liquid ice surfers: The construction of surfer identities in Norway. *Journal of Adventure Education and Outdoor Learning, 12*(1), 3–23.

Lawson-Remer, T. and Valderrama, A. (2010). Collective action and the rules of surfing. Retrieved 5 May 2015, from http://ssrn.com/abstract=1420122.

LeCompte, M. D. and Schensul, J. J. (1999). *Analyzing and interpreting ethnographic data*. Walnut Creek, CA: AltaMira Press.

Leonard, A. (2007). Learning to surf in Kuta, Bali. *Review of Indonesian and Malaysian Affairs, 41*(1), 3–32.

McHugh, P. (2003). Surfing's scary wave: 'Localism' intensifying at ocean breaks. *San Francisco Chronicle.* Retrieved 17 November 2016, from www.sfgate.com/cgi-bin/article.cgi?file=/chronicle/archive/2003/05/15/SP97503.DTL.

Mendez-Villanueva, A. and Bishop, D. (2005). Physiological aspects of surfboard riding performance. *Sports Medicine, 35*(1), 55–70.

Nazer, D. (2004). The tragicomedy of the surfers' commons. *Deakin Law Review, 9*(2), 655–713.

Nourbakhsh, T. A. (2008). *A qualitative exploration of female surfers: Recreation specialization, motivations, and perspectives.* (MSc), California Polytechnic State University, San Luis Obispo.

Olivier, S. (2012). 'Your wave, Bro!': Virtue ethics and surfing. *Sport in Society, 13*(7/8), 1223–1233.

Orams, M. B., and Towner, N. (2013). Riding the wave: History, definitions, and a proposed typology of surf-riding tourism. *Tourism in Marine Environments, 8*(4), 173–188.

Ormrod, J. (2005). *Endless Summer* (1964): Consuming waves and surfing the frontier. *Film & History, 35*(1), 39–52.

Pearson, K. (1982). Conflict, stereotypes and masculinity in Australian and New Zealand surfing. *The Australia and New Zealand Journal of Sociology, 18*(2), 117–135.

Pendleton, L. H. (2002). A preliminary study of the value of coastal tourism in Rincón, Puerto Rico. Los Angeles, CA.

Ponting, J. (2009). Projecting paradise: The surf media and the hermeneutic circle in surfing tourism. *Tourism Analysis, 14*(2), 175–185.

Ponting, J. (2014). Comparing modes of surf tourism delivery in the Maldives. *Annals of Tourism Research, 46,* 163–184.

Ponting, J., and O'Brien, D. (2014). Liberalizing Nirvana: An analysis of the consequences of common pool resource deregulation for the sustainability of Fiji's surf tourism industry. *Journal of Sustainable Tourism, 22*(3), 384–402.

Ponting, J. and O'Brien, D. (2015). Regulating 'Nirvana': Sustainable surf tourism in a climate of increasing regulation. *Sport Management Review, 18*(1), 99–110.

Preston-Whyte, R. (2002). Constructions of surfing space at Durban, South Africa. *Tourism Geographies, 4*(3), 307–328.

Rendle, E. J. and Rodwell, L. D. (2014). Artificial surf reefs: A preliminary assessment of the potential to enhance a coastal economy. *Marine Policy, 45*, 349–358.

Rider, R. (1998). Hangin' ten: The common-pool resource problem of surfing. *Public Choice, 97*, 49–64.

Scheibel, D. (1995). 'Making waves' with Burke: Surf Nazi culture and the rhetoric of localism. *Western Journal of Communication, 59*(4), 253–269.

Scott, P. (2003). 'We shall fight on the seas and the oceans … we shall': Commodification, localism and violence. *M/C: A Journal of Media and Culture, 6*(1).

Sotomayor, S. and Barbieri, C. (2015). An exploratory examination of serious surfers: Implications for the surf tourism industry. *International Journal of Tourism Research.* doi: 10.1002/jtr.2033.

Stranger, M. (2011). *Surfing life: Surface, substructure and the commodification of the sublime.* Farnham, UK: Ashgate.

Surfing for Autism. (2015). About us. Retrieved 6 June 2015, from www.surfingfor autism.org/about-us/.

Sweeney, S. H. (2005). *The spatial behavior of surfers.* University of California at Santa Barbara Department of Geography. Santa Barbara, CA.

United States Surfing Federation. (n.d.). Wounded Warrior Project. Retrieved 6 June 2015, from www.surfusa.org/contact/.

Usher, L. and Calhoun, S. (2015). *Claiming Pavones: Expatriate surfers and territoriality in Costa Rica.* Paper presented at the International Congress on Coastal and Marine Tourism, Kailua-Kona, HI.

Usher, L.E. and Gómez, E. (2016). Surf localism in Costa Rica: Exploring territoriality among Costa Rican and foreign resident surfers. *Journal of Sport & Tourism.* doi: 10.1080/14775085.2016.1164068.

Usher, L. E. and Kerstetter, D. (2015a). Re-defining localism: An ethnography of human territoriality in the surf. *International Journal of Tourism Anthropology,* in press.

Usher, L. E. and Kerstetter, D. (2015b). 'Surfistas locales': Transnationalism and the construction of surfer identity in Nicaragua. *Journal of Sport and Social Issues.* doi: 10.1177/0193723515570674.

Waitt, G. (2008). 'Killing waves': surfing, space and gender. *Social & Cultural Geography, 9*(1), 75–94.

Waitt, G. and Frazer, R. (2012). 'The vibe' and 'the glide': Surfing through the voices of longboarders. *Journal of Australian Studies, 36*(3), 327–343.

Waitt, G. and Warren, A. (2008). 'Talking shit over a brew after a good session with your mates': Surfing, space and masculinity. *Australian Geographer, 39*(3), 353–365.

Walker, I. H. (2011). *Waves of resistance: Surfing and history in twentieth-century Hawai'i.* Honolulu, HI: University of Hawai'i Press.

Warshaw, M. (2003). *The encyclopedia of surfing.* Orlando, FL: Harcourt.

Weisberg, Z. (2010). Rivas Province, Nicaragua. *Surfer, 51*, 108–110.

Wheaton, B. (2007). Identity, politics and the beach: Environmental activism in Surfers Against Sewage. *Leisure Studies, 26*(3), 279–302.

Young, N. (ed.). (2000). *Surf rage: A surfer's guide to turning negatives into positives.* Angourie, NSW: Nymboida Press.

10 Spot X

Surfing, remote destinations and sustaining wilderness surfing experiences

Mark Orams

Introduction

The global spread and evolution of surfing as a recreational and sporting activity has been pervasive and rapid. From its introduction to a limited range of locations during the 1950s through the following half-century, surf-riding has become a massive global activity which involves millions of participants riding waves in a multitude of ways across tens of thousands of locations. Furthermore, the act of riding waves is only a relatively minor aspect of the influence of the activity of surfing. Its effects extend into culture, language, music, art, movies, tourism, clothing, vehicles, coastal development and lifestyles (Orams and Towner, 2013). The seduction of and addiction to surfing is, therefore, a widespread and influential human phenomenon.

The spread of surfing has become the catalyst for the development of coastal locations. The discovery of a high-quality surf break previously unknown, or little known (or perhaps not surfed before) has become a highly prized pursuit for many surfers. The sense of exploration, the adventure and excitement of the unknown, and the reward/satisfaction of being the first or pioneer for surfing new breaks is all part of the attraction. This approach has been reflected in and reinforced by surfing-related companies marketing and branding strategies (e.g. Rip Curl's 'The Search') and, particularly by the surf-movie genre where numerous story lines are based around following the adventure of surfers as they seek out new, previously unknown surfing breaks (e.g. *The Endless Summer*, *The Green Iguana*). These discoveries and the subsequent publicity and profile for the surf break almost always leads to others visiting the locations to surf and consequently a growth in surfer numbers at the break occurs. Associated infrastructure and services develop as both visitors and locals recognize an opportunity to meet the needs/wants of the growing number of visiting surfers.

Such a pattern of discovery, growth and in a number of cases crowding, conflict and environmental degradation and a diminished quality of surf experience causes many to lament or actively oppose this surf break location development tendency. One manifestation of this is the 'locals only' catch-cry, whereby local surfers assert an informal ownership and control and oppose, restrict or shape the visitation or behaviour of surfers from elsewhere (Daskalos, 2007; Rider,

1998). This is a cause of conflict at many surfing locations (Bandeira, 2014). Another approach has been the deliberate restriction of information about or keeping secret the location of high-quality surf breaks that have been discovered. In many surfing magazines, websites, movies and other public forums surfers refuse to disclose the details of the location where images of high-quality waves and stories of special surfing experiences are shared. Thus the generic descriptor of 'Spot X' attached to such images/stories to protect the spread of information about the surf break.

Both of these approaches are derived from perceptions regarding the quality of the surfing experience. For most of surfing history, an informal rule of one surfer per wave has persisted, so much so that protocols for surfing extend across cultures and locations so that the surfer who catches a wave first and rides closest to the breaking wave peak (where the wave is steepest) is deemed to have the right of way and exclusive rights to surf that wave alone. Any other surfer attempting to catch this same wave in the same part of the wave is deemed to be 'dropping in' and breaking the protocol. Such transgressions are frequently aggressively enforced with verbal abuse, threats, physical confrontations and, on some occasions, violence. The 'one wave–one rider' protocol is so widely understood, accepted and reinforced it is seldom explored or questioned. Interestingly, the originators of modern surfing in Hawai'i (whilst acknowledging that many indigenous peoples elsewhere in the world have ridden waves for centuries) did not appear to have this exclusivity (one surfer per wave) as part of their practice. Photographs and written accounts of surfing in Hawai'i almost always show multiple surfers and surfing craft sharing the same wave. For many recreational casual surfing locations where there are large numbers of surfers and surf-craft, wide open beaches and small crumbling waves there is widespread acceptance of sharing waves. However, at high-quality surf breaks, even when crowded, sharing of waves is not accepted and at professional surfing contests the one surfer per wave protocol is formally built into their rules (with penalties for surfers who transgress the 'priority' rule).

The one surfer per wave protocol is likely derived from several issues, some pragmatic and others related to the quality of the experience. First, surfing large challenging waves can be dangerous and another surfer on the same wave can have safety implications. They can be unpredictable in terms of where they will manoeuvre on the wave, they may fall ('wipe-out') and their surfboard (or other surf-craft) or their body could hit the other surfer. Second, surfers derive enjoyment (and are admired by others) for their ability to conduct and successfully complete manoeuvers on the wave. These manoeuvers are usually undertaken on or close to the critical section of the wave closest to the breaking peak and or on the open (or tubing) smooth face of the wave. Thus, there is a highly desired part of the wave and space needed around that to manoeuvre and complete 'cut-backs, re-entries, floaters, aerials, tube/barrel rides, bottom turns, 360s' and so on. Finally, the sense of alone-ness on a wave, of being at one with the wave, connected to it, part of it, moving with it is a deeply personal one and deeply ingrained in most surfer's being. This description is an attempt to give insight into the cause of the addiction to surfing so many surfers feel. It is captured in another surf-clothing company's slogan 'only a surfer knows the feeling', which reflects that for many (perhaps most) surfers their connection to and passion for surfing is a deeply felt one that goes beyond competitive success, personal enjoyment, kudos from others, image or a fun activity. Rather it touches something within them that is not able to be explained (at least not easily) or quantified. It is part of who they are (as opposed to something they do).

In summary therefore, surf-riding is a global phenomenon that has developed and spread rapidly over the past 50 years. The quality and safety of surf-riding, in many circumstances, is negatively affected by large numbers of surfers (crowds) and, as a consequence, un-crowded waves and surfing locations are highly valued and sought after. There is a tension therefore, between the growth of surf-riding in terms of its increasing popularity and the quality of the surfing experience (at least as perceived by many surfers).

Relevant tourism development models and concepts

There are numerous case studies whereby tourism has been the catalyst for economic development in remote locations. Typically, such development occurs as a consequence of the recognition of a specific attraction desired by outsiders for which they are prepared to travel and invest money to visit and experience. Within the tourism literature there are a number of well-established (and tested and critiqued) models which have relevance for considering the implications of surf-riding tourism development.

Butler's tourism destination life-cycle

The concept of a product life-cycle is well known in business and marketing literature (Rink and Swan, 1979), and this concept was the basis for the proposal of a destination life-cycle first put forward by Richard Butler (1980). Fundamentally the model outlines a typical transition from a discovery phase, to a growth phase to an eventual plateau or atrophy phase over time. When applied to a tourism destination it points to a range of changes that occur at the destination as tourism development becomes the catalyst for a range of changes economically, socially and environmentally. The model has been widely used as a basis for understanding (and even predicting) general patterns of change that occur as a consequence of tourism development (Orams, 1999).

Butler's model has relevance for surfing destinations where numerous examples exist of a pattern of development that has occurred over time as a location becomes known and attractive for surfing and the subsequent changes that occur. An examination of the history of now well-established surfing destinations such as Jeffrey's Bay, South Africa, Uluwatu and Kuta Beach, Bali, Indonesia, the North Shore of Oahu, Hawai'i, Surfer's Paradise, Queensland, Australia, Malibu Beach, California, Newquay, Great Britain reveals a massive transition from the 1960s to the turn of the twenty-first century. Each of these locations was once a relatively lowly populated coastal area without any surfing activity. However, at some point (mostly in the 1960s) pioneering surfers tested out and began to ride waves at the breaks and this proved the catalyst for additional surfers and growth in popularity, profile and associated development of the nearby coastal area. In the great majority of cases, the growth in numbers of surfers and the commensurate development in the associated coastal locale have followed the pattern that Butler's destination life-cycle proposes. Where this has not been the case, the surf breaks have been offshore or only accessible via boat or where the break is so remote that land-based access is very difficult.

Doxey's Irridex

The growth of a tourism destination is based on a growing number of visitors or outsiders spending time within a place that is not their primary place of residence. An increasing influx of outsiders into a community has a range of impacts and effects on residents. Doxey (1975) proposed that there could be a pattern of general reaction from locals with increasing numbers of visitors (tourists) over time. He called his model the Irritation Index or 'Irridex' and argued that local communities moved from a position of being open, welcoming and positive about visitors to a point, over time, where they become increasingly irritated or negative (or they adapt and there is acceptance or apathy) about the number of visitors and their influence. While the model is by necessity generic and will not reflect the views or experiences of all locals or all tourism destinations it does emphasize an important point: that tourism development has an influence on local people. Furthermore, that tourism development is not viewed universally as a positive phenomenon.

A similar phenomenon could be postulated to occur with regard to surfers' reaction to growth. More specifically, that in early 'discovery' days, when surfers are rare visitors, the reaction to the arrival of other surfers is generally positive, welcoming and adds value to the experiences of the surfers (through the sharing of experiences on and off the water). However, as the surf break grows in popularity it eventually reaches a stage where the numbers of surfers arriving and/or living locally becomes so large that the crowding of the surf breaks produces negative reactions, irritation, antagonism and even aggression.

Recreational succession

Orams (1999) applied the term recreational succession from its origins in the US Forest Service (Stankey, 1985) to describe the pattern of transition that can occur when a new coastal or marine attraction is discovered and subsequent development of the site as an increasingly popular location for recreation and tourism. With increasing popularity and development the features that attracted the pioneers or discoverers are lost (for example, pristine natural settings, wilderness, little contact with other visitors, 'authentic' interactions with locals) and consequently this group of visitors are displaced.

As pristine natural sites are discovered and used for recreation, deterioration of the site's natural attribute occurs. Consequently, initial visitors, who were attracted by the pristine unspoiled surrounds, move on and are replaced by greater numbers, with lower expectations of environmental quality. This chain continues, resulting in ever increasing numbers of visitors, increasing development of the site's infrastructure to cope with visitors' needs and decreasing environmental quality. Meantime, the initial discovering group, having moved on, have explored and discovered another pristine site and thus have started the chain of recreational succession again elsewhere.

(Orams, 1999, p. 59)

Thus, recreational succession is correlated with the pattern of development identified in Butler's destination life-cycle model and is derived from the reaction of locals and visitors as tourism grows as postulated by Doxey's Irridex. What recreational succession identifies that adds to these models is the influential role that explorers and pioneers play in discovering new locations and providing the catalyst for similar patterns of development at multiple locations.

Displacement

Related to the concept of recreational succession is the displacement of both local residents and visitors to locations as they become more popular. This trend has been reported in a variety of nature-based recreation and tourism settings where crowding and related development and changes in the quality of the natural resources proves a 'push' factor whereby people decide to move away or no longer visit (e.g. Ednie and Leahy, 2008; Kearsley and Coughlan, 1999). In addition, as locations become more popular for tourism, local residents may find the cost of living too high and be forced away or conversely, find that the value of their property increases to a point where they decide to sell and utilize the capital gain to re-establish themselves elsewhere.

The relevance to surfing

Because the quality of the surfing experience is affected by crowding (Ford and Brown, 2006; Orams and Towner, 2013) and also, for many surfers, the quality of the natural setting where a surf break is situated (Ponting, 2009), the development and popularity of a surf break and the surrounding area is an influential issue. In addition, the draw of exploring, discovering and surfing previously unknown surfing breaks is enormously seductive. As a consequence, there are a huge number of surfers who are prepared to spend money, time and effort to travel to discover new or rarely surfed breaks.

The irony of the focus on exploring and discovering new surfing destinations is that in doing so, these explorers can unintentionally become the genesis of the very experience and outcome they detest and are seeking to escape. Namely, a crowded surf break and a human modified and commercialized local coastal town or city (or surfing-focused resort).

The concept and value of wilderness

The concept of wilderness has its origins in the beginnings of the modern day conservation movement which led to the creation of protected natural areas such as national parks (Orams, 1999). While there are a variety of definitions, essentially wilderness refers to large areas of high-quality natural ecosystems where there is little or no human influence and where opportunities for recreational interaction with nature include experiencing solitude, freedom and self-sufficiency.

The benefits of designated wilderness areas for both environmental conservation and in terms of the opportunities for quality recreation experiences have long been advocated (Hendee *et al.*, 1978). In particular, research has identified the value of wilderness experiences for self-sufficiency, skill acquisition, self-confidence, personal resilience, mental health, social interaction and relationships and awareness of and support for nature (and its conservation) (Cole, 2004). While the majority of this work has focused on terrestrial settings, especially mountains, forests and deserts, some attention is being given to coastal and marine settings as significant wilderness settings (Orams, 1999). Interestingly, Aldo Leopold (1949), one of the foremost early writers advocating for the importance of protecting wilderness identified in the early twentieth century that coastal locations were most at risk in terms of development and loss of wilderness:

> One of the fastest-shrinking categories of wilderness is coastlines. Cottages and tourist roads have all but annihilated wild coasts.... No single kind of wilderness is more intimately woven with history, and none nearer the point of complete disappearance.
>
> (Leopold 1949, p. 194)

This historical and long-standing work on the importance of conserving wilderness (and the formal designation and protection of wilderness areas) and the experiences they offer has relevance to surfing. For many surfers, the desire to travel to remote surfing locations is driven by a desire to escape crowds but also by a desire to experience the opportunities provided by a wilderness setting for their surfing, particularly adventure (Reynolds and Hritz, 2012).

Diminishing opportunities for wilderness

As the global human population continues to grow, the number of locations on the planet where humans have little influence diminishes. This is particularly the case with regard to coastal areas where high-quality recreational experiences are available (as Leopold identified in the mid twentieth century). The demand for coastal locations and the value they hold for humans is reflected in the inflated property prices of coastal locations when compared with similar inland locations (Peart, 2009). Places of human settlement, particularly urban areas tend to 'sprawl' along the coast. Furthermore, coastal settings, particularly sandy beaches, are the most popular for second homes and as venues for vacations (Orams, 2003). Thus, coastal settings and near-shore coastal environments that are true wilderness are increasingly rare. This is recognized as a continuing challenge for the twenty-first century even pertaining to remote open ocean ecosystems (Graham and McClanahan, 2013). Because surf breaks depend on a specific combination of natural conditions (ocean swell, bottom geomorphology, depth/tide and wind) for high-quality surfing conditions their number is limited globally. Furthermore, as the popularity of surf-riding in all its diverse types continues to grow, high-quality surf breaks are becoming more highly sought after, and an associated industry of surf-riding tourism has developed (Orams and Towner, 2013). This in turn provides incentives for coastal development based around accessing these surf breaks to occur.

Responses to diminishing wilderness, increasing popularity of surfing and crowding

The need to protect surf breaks, wilderness and opportunities for high-quality surfing experiences is becoming recognized as a significant challenge and a number of advocacy groups have been established to lobby for greater protection of high-quality surf breaks. Examples include: the Surfrider Foundation, which was founded in 1984 with a mission for the 'protection and enjoyment of the world's ocean, waves and beaches through a powerful activist network' (www.surfrider.org), the Surfbreak Protection Society in New Zealand (www.surfbreak.org.nz) and Surfers Against Sewage (www.sas.org.uk). In addition, a number of authors have been advocating for the extension and application of the parks and reserves concept to surfing breaks (e.g. Farmer and Short, 2007; Short and Farmer, 2012).

Additional approaches include attempts to restrict the number of surfers in the water at specific breaks. This occurs either informally, via aggression from local surfers (i.e. surfers who live locally or have a long-standing association with a particular surf break) against visitors or newcomers (e.g. 'locals only' edicts) (De Alessi, 2009) or formally, via restricting the number of surfers permitted in the water at specific breaks at any one time. Other less formal aspects of the wider surfing culture is for surfers to keep surf breaks secret or to limit information about the location and ideal conditions for specific breaks to themselves, hence the 'Spot X' moniker for photographs and stories about high-quality surfing in magazines, online and other publically available forums.

There is, therefore, increasing recognition of the importance of providing protection of the existence and functioning (i.e. in terms of the conditions that are conducive to high-quality surfing waves) of surf breaks and the need to also control development so that wilderness values are protected. However, this growing recognition exists within the broader context of a growing global human population, an increasing number of surfers and diversity of surf-riding activities (Orams and Towner, 2013) and a global economic system which is based on an almost universal focus on continued growth. Herein, therefore, lies the dilemma. On one side, a growing demand for high-quality surf breaks, and economic and political systems that are based on growth as a desired and important aspiration and on the other side a limited (or perhaps decreasing) number of high-quality wilderness surf breaks.

The future: challenges and possibilities

The data on international tourism show a continuing general trend of growth and with the emergence of increased personal wealth for large cohorts of developing nations' populations (e.g. China and India) (UNWTO, 2014), all predictions are for international tourism to continue to grow over the next decade and beyond (UNWTO, 2014). Furthermore, while a substantive amount of this growth is in the traditional mass tourism market to well-established destinations, there is a significant growth in niche and new tourism destinations, including remote islands (UNWTO, 2014). In addition to this wider growth in international tourism, surf-riding activities are also growing in popularity and diversity. Travel for surfing is a fast-growing aspect of the tourism industry.

Modern media and popular culture also play an increasingly influential role in promoting growth to non-traditional tourism destinations. There is no doubt that the desire to travel to specific locations to surf is predicated by an awareness of the opportunity to surf specific surf breaks. If the surf break is not known, then it is not surfed (unless it is discovered by accident or a deliberate exploration of an area of coast-line with the intention of finding a new surf break). Social media, swell/wave/surf/weather prediction models, improved navigation (including the ability to search for potential breaks remotely via tools such as Google Earth) are all contributing to the growth of surf-riding tourism. Thus, surf-riding tourism is set to continue to grow and the seeking out of new, remote and high-quality waves to ride will become even more widespread and influential. Related to this growing demand are a number of responses and opportunities which are being developed and which add to the spectrum of surf-riding opportunities.

Technology is also influential in the growth surf-riding tourism. A wider range of tools, toys and possibilities are now available than ever before. These provide access to previously unknown or inaccessible activities associated with surfing. Examples include in-water-based photography and videography (such as that available via GoPro and other similar image capture devices), foiling surf-boards, personal-water-craft (and tow-in surfing opportunities), drones for photography and exploring new breaks, stand-up paddle surfboards, kite-boards (including foiling boards), warmer, more flexible wetsuits, dry-suits, tide and weather prediction watches.

While technology is influencing surfing activities it is also having an influence on the supply-side as well. Over the past two decades a number of attempts have been made to create artificial (human made) reefs and enhanced surf breaks (Gough, 1999). In addition, the design and building of land-based wave-pools and human made surf breaks as specific recreation/tourism attractions has become more widespread.

These developments are likely to continue as creativity, entrepreneurship and design engineering seeks to take advantage of the demand for surf-riding. It is likely that virtual-reality based surfing experiences are not far away. Such continued developments may provide both a counter to the diminishing wilderness surfing opportunities (in that they could provide an alternative) or as an additional contribution to the rising demand (as a catalyst for more to want to pursue real 'natural' surfing).

The implications of a system based on continual growth

Because the world's economic system is based on an underlying objective of growth, the implications are global. While acknowledging some recent advocacy for a change in this approach, for example see the work on steady state economics (Daly, 1991), and alternative measures of success other than GDP growth (van den Bergh and Antal, 2014), the pervasiveness and unquestioning acceptance of this objective is surprising. From an individual to a global level, ambitions to grow and develop into having more access to resources, financial wealth, material possessions and 'standard of living' are dominant. Despite decades-long attention being given to transitioning to a more sustainable form of living (even this was couched in the terms of 'sustainable development') little progress has been made in this regard.

This trend has implications for all life, not only human activity, and surfing and the development of surf-riding destinations are not excluded. As human population continues to grow and as the number of surfers continues to grow, the demand for surfing opportunities will also grow. The consequence of this is that increased crowding, further development and diminishing wilderness surfing experiences will continue. The challenge is obvious, but perhaps the opportunities less so. Because there is an increasing recognition of the threat posed by these growth trends, there is also a commensurate recognition of the need to protect surf breaks and surfing experiences (including wilderness). Hence the advocacy work of the Surfrider Foundation, Surfers Against Sewage, authors (this volume for example) and research/educational collaborations (e.g. The Centre for Surf Research: www.centerforsurfresearch.org) and others who are pushing for formal protection of surf breaks and aspects that are important influences on surfing experiences such as clean water, healthy marine ecosystems and protection of near-shore coastal habitats. In addition, there are a growing number of organizations being established that seek to shape surfing and surf-riding tourism into a force for positive social and environmental change through philanthropy, entrepreneurship, volunteerism and science (e.g. SurfAid

International: www.surfaid.org, Waves for Development: www.wavesfor development.org, Project Wave of Optimism – Project WOO: www.projectwoo. org, SurfCredits: www.surfcredits.org, Surf Resource Network: www.surfresource. org, European Association of Surfing Doctors: www.surfingdoctorseurope.com, Waves of Freedom: www.wavesoffreedom.org).

The need for coastal and marine wilderness protection as a priority for the future

So, while the challenges associated with further growth are significant, there is also a growing community who seek to shape surfing and its influence in a positive way for the future. A critical contributor to this will be the formal designation of areas of coastline as wilderness resources that protect the diminishing number of locations where there is little or no human influence. Coastal and marine wilderness areas of a size and significance which mirrors the enlightened approach shown in the setting aside of large areas of land, forests, mountains, lakes and rivers as national parks and wilderness areas in the nineteenth and twentieth centuries is needed. The establishment of mega-marine protected areas, which has achieved some notable success in the early part of the twenty-first century, is encouraging and is to be applauded. It needs to continue and surfers and all ocean lovers need to keep working to ensure this legacy is expanded. The future for surfing and the opportunity for enjoying it in natural settings will depend on the ability of the surfing community to ensure its desires are accounted for in resource management decision making. Furthermore, both the advocacy for further protection of coastal wilderness areas and action to restore coastal areas that have been compromised through human activity, influence or natural disaster will be needed if wilderness surfing experiences are not to become a rarity.

References

Bandeira, M. M. (2014). Territorial disputes, identity conflicts, and violence in surfing. *Motriz: Revista de Educação Física, 20*(1), 16–25.

Butler, R. W. (1980). The concept of a tourist area cycle of evolution: Implications for management of resources. *Canadian Geographer, 24*(1), 5–12.

Cole, D. N. (2004). Wilderness experiences. *International Journal of Wilderness, 10*(3), 25.

Daly, H. (1991). *Steady-state economics*. Washington, DC: Island Press.

Daskalos, C. T. (2007). Locals only! The impact of modernity on a local surfing context. *Sociological Perspectives, 50*(1), 155–173.

De Alessi, M. 2009. The customs and culture of surfing, and an opportunity for a new territorialism? *Reef Journal, 1*(1): 85–92.

Doxey, G. V. (1975). A causation theory of visitor-resident irritants: Methodology and research inferences. In *Conference Proceedings: Sixth Annual Conference of Travel Research Association*. San Diego, CA, pp. 195–198.

Ednie, A. and Leahy, J. (2008). Recreation succession. In *The encyclopedia of tourism and recreation in marine environments*, Lueck, M. (ed.) (p. 391). Wallingford, UK: CABI.

Farmer, B. and Short, A. D. (2007). Australian national surfing reserves: Rationale and process for recognising iconic surfing locations. *Journal of Coastal Research*, *50*(SI), 99–103.

Ford, N. and Brown, D. (2006). *Surfing and social theory: Experience, embodiment and narrative of the dream glide*. Abingdon, UK: Routledge.

Gough, V. J. (1999). *Assessing the economic effects of recreation facility development: Proposed artificial surfing reef, Mount Maunganui, New Zealand*. Unpublished Honours thesis, University of Waikato, Hamilton.

Graham, N. A. and McClanahan, T. R. (2013). The last call for marine wilderness? *Bioscience*, *63*(5), 397–402.

Hendee, J. C., Stankey, G. H. and Lucas, R. C. (1978). *Wilderness management* (No. 1365). Forest Service, US Department of Agriculture.

Kearsley, G., and Coughlan, D. (1999). Coping with crowding: Tourist displacement in the New Zealand backcountry. *Current Issues in Tourism*, *2*(2–3), 197–210.

Leopold, A. (1949). *A sand county almanac*. New York: Oxford University Press.

Orams, M. B. (1999). *Marine tourism: Development, impacts and management*. London: Routledge.

Orams, M. B. (2003). Beaches as a tourism attraction: A management challenge for the 21st century. *Journal of Coastal Research* (special issue), *35*, 74–84.

Orams, M. B. and Towner, N. (2013). Riding the wave: History, definitions, and a proposed typology of surf-riding tourism. *Tourism in Marine Environments*, *8*(4), 173–188.

Peart, R. (2009). *Castles in the sand: What's happening to the New Zealand coast?* Nelson, New Zealand: Craig Potton Publishing.

Ponting, J. (2009). Projecting paradise: The surf media and the hermeneutic circle in surfing tourism. *Tourism Analysis*, *14*(2), 175–185.

Reynolds, Z. and Hritz, N. M. (2012). Surfing as adventure travel: Motivations and life-styles. *Journal of Tourism Insights*, *3*(1), 1–17.

Rider, R. (1998). Hangin'ten: The common-pool resource problem of surfing. *Public Choice*, *97*(1–2), 49–64.

Rink, D. R. and Swan, J. E. (1979). Product life cycle research: A literature review. *Journal of Business Research*, *7*(3), 219–242.

Short, A. D. and Farmer, B. (2012). Surfing reserves: Recognition for the world's surfing breaks. *Reef Journal*, *2*, 1–14.

Stankey, G. (1985). *Carrying capacity in recreational planning: An alternative approach*. United States Department of Agriculture – Forest Service. Ogden, UT.

UNWTO. (2014). *UNWTO Tourism Highlights, 2014 Edition*. Madrid: United Nations World Tourism Organization. Retrieved 25 November 2016, from http://dtxtq4w60xqpw.cloudfront.net/sites/all/files/pdf/unwto_highlights14_en.pdf.

Van den Berg, J. C. J. M. and Antal, M. (2014). *Evaluating alternatives to GDP as measures of social welfare/progress*. Working Paper no 56. WWW for Europe. Retrieved 10 June 2015 from www.wifo.ac.at.

11 Surfing

A ritual with consequences

Jon Anderson

Introduction: chasing utopia

It was the day after Valentine's Day and I was spending some time back home in Canada doing some much needed editing work at my office ... only to hear that familiar 'pop' in my web browser. It was Facebook chat. The message was from a friend of mine in Morocco. The cryptic message read 'Thursday is going to be the day'.... I quickly opened the surf forecast model on my browser and animated the North Atlantic. A massive low pressure was headed on a collision course with Ireland.... Seas upward of 50 feet and wave intervals in the 19 second range.... It was too late for me to chase this swell there. The best option would be a thousand kilometres south, in Africa and I still had time to make it. Only catch – I had three hours to make the flight.

...As I pulled into the airport, I thought how our technology is changing..... A few years ago it would have been impossible to book a ticket through an iPhone on the way to the airport, let alone answer emails and look at surf forecasts.... As I parked my car in the long term parking lot and pulled out my bags ... I said out loud to myself, 'Too easy' with a big smile on my face.

...The next few days felt like a Utopic Groundhog day where the same conditions would repeat themselves. Upon returning ... back home in Canada ... suddenly this all too familiar noise popped up on my computer. It was Facebook chat. It was my friend just up the shore. He informed me this fabled right point break was absolutely firing and doing an impersonation of Rincon sprinkled with icing sugar.... I got in my car and headed for the coast....

(Ouhilal, 2011: 88/95/97)

For surfers, the lure of swell is tantalising. Regardless of their chosen means of wave riding (using body, body-board, short- or long-board, ski, sit on top, paddleboard or kayak), every surfer dreams of the breaking wave. With the advent of new technologies such as online wave forecasting, credit cards and social media, alongside the affordability of new transport systems including commercial jet travel, surfers are able to respond to the lure of swell wherever in the world it occurs. As Ouhilal's case exemplifies (above), many surfers are able to enjoy a lifestyle that is defined by the changing geography of surfing. Surfers chase the utopia of the tube.

> As the World shrinks and frontiers retreat in the face of a frenzied search for perfect, uncrowded surf, [we should] take a moment to reflect on what it means to be a global surf traveller.
>
> (Colas, 2001: 4)

Chasing the place of the tube raises questions for the sustainability of this life-style sport. In light of the broader aims of this volume, the chapter will argue that surf culture needs to reflect seriously on what it means to be a global surf traveller (after Colas, above). According to big wave rider Laird Hamilton, surfing lives are defined by 'ritual[s] of consequence' (Casey, 2011: 376). In line with the groundswell of writings in surf tourism literature (see for example, Buckley, 2002; Ponting and McDonald, 2013; Barbieri and Sotomayor, 2013; Ponting, 2014; West, 2014; Ponting and O'Brien, 2015; and Martin and Assenov, 2012 for a point in time overview), in this chapter I argue that surfing and surf travel are, for its participants, rituals *with* consequence. This chapter makes the case that surf travel has significance for the surfer involved in the activity – producing healthy bodies, emotional affects, skill development, social status and, for some, wealth and fame; whilst it also has significance beyond the participant, for the cultural places hosting surf locations, and the broader environment which contends with the pollutants produced from human (in this case surfer) activities. However, it is often the case that in surf culture the internal consequences of the surfing life dominate discussion and interest; the strengths of this volume lie in how the broader external consequences of surfing are raised and debated. In this chapter, I explore the externalities produced by surfing in respect to the two key drivers of surfing activity noted above: the place of the tube and the practice of travel. In terms of the practice of travel, the chapter raises the issue of the carbon costs involved in commercial air travel; and in terms of the place of the tube, the chapter outlines the environmental, cul-tural and social costs locations suffer through the influx of surfers into an area. The chapter will focus in particular on the case of Bali and Gili Trewangen in Indonesia. Using ethnographic accounts and photos from the author, alongside secondary materials from surf media and Indonesian articles, the chapter uses this case to question whether the 'utopic Groundhog days' enjoyed by Ouhilal (above) should be simply framed as internally pleasurable events, or whether they must be contextualised as rituals with external consequences too. The chapter suggests that the debate over the nature of surfing utopia could be use-fully informed by lessons from environmentalism, including a new 'coyote' pol-itics of pragmatism that enables the integration of surfing with sustainability more strongly. This approach involves identifying and accepting the external consequences of surfing practice, and committing to incrementally integrate positive changes that aim for a moral and responsible future for the ritual of surfing.

The surfing ritual

Surfing lifestyles are defined by the place of the tube and the practice of travel to get there. With respect to the place of the tube, surfers surf in order to put themselves in direct association with the energy of cresting waves, the latter driven by broader oceanic and atmospheric systems and formed through the meeting of water, air and specific geological locations. Becoming part of this elemental convergence (see Anderson, 2012a) produces specific internal consequences for the surfer. Often articulated in terms of the numinous (Otto, 1970; Graber, 1976), sublime (Burke, 1776; Stranger, 1999), or flow (Csíkszentmihályi, 1990), surfers themselves commonly refer to the relational sensibility generated between themselves and the elements as 'stoke' (see Anderson, 2014a); a feeling which plugs the power of the ocean 'into [a surfers'] adrenal glands' (McGinty, cited in Warshaw, 2004: xx). The affective rush gained from the place of the tube is powerful and perhaps even addictive, and generates an urge to experience new waves at new locations (see Doherty, 2007). This urge is so strong that it has driven the expansion of the surf tourism industry. Surf tourism has become 'a rapidly expanding market segment of the wider sport tourism industry' (Martin and Assenov, 2012: 257), and as West points out, over recent years surf tourism has grown into not only, 'one of the most robust sectors of the tourism industry, [but also] the economic linchpin for the global surf industry' (2014: 412). The surf tourism market is estimated to be at least one-quarter of a billion US dollars (Ponting, 2008), and features multiple forms of tourism, including high-end resort tourism and ecotourism. Surf tourism is driven by surfers' anticipation and excitement of riding waves in different places; the challenges posed by new waves and conditions transfers a love for travel and exploration to many surfers' lives. Surfers' urge to travel the planet is not simply for its own sake (i.e. to keep moving from one random place to another), but rather to realise the relational sensibilities, skills and social status produced by communing with new iterations of elemental convergence.

Surfers' identity is therefore transformed by both the process of travel and the adrenalin hit achieved through the practice of riding waves in new water worlds (Anderson and Peters, 2014). It is these internal consequences of surfing that appear to preoccupy active participants in the sport, and the broader surfing culture that surrounds it. It is questionable whether the external consequences of these identity practices are reflected on in depth within this culture, and it is with this problematic that the language and practice of sustainability must engage. As I have noted elsewhere (Anderson, 2010), sustainability explicitly seeks to negotiate a middle ground between the often competing claims of the economy, the environment and society (not to mention future generations) (World Commission on Environment and Development, 1987). However, even as the most well-known attempt to offer a way forward for the environmentalist message, the language of sustainability and sustainable development (SD) has failed to engage many sectors of wider society (see Seghezzo, 2009). As Macnaughten and Jacobs (1997) state in relation to discussions with the general public on the subject,

> It is tempting to ask whether any single definition the public could provide when challenged to explain 'SD' would be considered correct.... When introduced into the discussion by the moderator, the term was generally seen as a piece of abstract jargon; even 'gobbledygook'.
>
> (1997: 15)

So as we begin to consider what it means to be a global surf traveller we should perhaps first reflect upon two stark questions: first, why *should* sustainability be part of any surfing utopia? And second, why should *surfing culture* be any more willing to engage with this 'unwieldy' concept in their daily lives than any other social group (Darnton, 2004: 8)? To the lay surfer, sustainability could be seen to refer to the ability of waves to sustain identities and lifestyles. If taken on these terms, the place of the tube is a highly sustainable resource; waves will always generate stoke for surfers, and even when waves are not accessible, the energy gained from previous surfing sessions can sustain surfers during fallow periods of waiting (Ford and Brown, 2006). These stark questions are particularly resonant for surf culture in the developed world as here there are numerous individuals privileged enough to arrange their travel at the tap of a touchscreen. Developed world surf culture also celebrates the characteristics of hedonism, individuality and even narcissism (see Wheaton, 2005; Evers, 2010), traits which at first glance do not appear to synergise with the egalitarian and co-operative principles of environmentalist concerns. For many within surfing culture, therefore, the world's coasts are seen as their playground; an open, placeless space across which they can be transported without impediment or barrier (after Olwig, in Ingold, 2007: 78). This utopia of movement and geography is enjoyed as if it were immune from the consequences it produces for communities and host destinations. The surfer in this framing acts as a discrete and self-directed individual. As we have seen in the case above, for example, Ouhilal does not question *whether* he should travel to a particular destination; he does not appear to reflect on the environmental consequences of his travel, or the social and cultural consequences that his temporary presence in a foreign location may have for those that live (and potentially surf) there. Similarly, the economic costs, employment repercussions, or any family implications of his absence are not shared with the reader. By implication, it appears to be the case that as technology enables the surfer to overcome informational, social, environmental and geographical barriers, it is irrelevant for the individual to consider them. In this view, as surf-writer Clifton Evers puts it, surfers are individuals preoccupied by, ' "*our*" adventure ... surfing the waves, while *other* stories about the lives of local people and communities tend to be silenced. *It's all about us*' (2010: 22, my emphasis).

However, as this volume suggests, 'good' and 'responsible' surfing needs to be more than 'about us'; it needs to include the external consequences of surfing too. These consequences come in many different forms, in terms of both the place of the wave, and the practice of travel to get there. The next section explores these external consequences of the ritual of surfing with respect to one key global surf destination, Bali and the Gili islands, Indonesia.

'A place that cannot be?' (Ouhilal, 2011)

In line with this chapter, and the broader volume thus far, Ponting *et al.* have argued how surfing utopia can and should be deconstructed (2005, see also Ponting and McDonald, 2013). In their view, surfing wonderlands are versions of utopia that simply cannot exist; they are inappropriately framed as 'surfing tourist space[s] based upon adventure, the search for the perfect wave, uncrowded breaks and absent or compliant local communities' (Ponting *et al.*, 2005: 141). These constructions crucially 'silence' the 'other stories ... of local people and communities' that publicise the negative effects of travelling surfers' presence. As surfers travel to their dream waves, their means of mobility and on-shore presence creates environmental pollution, attracts the proliferation of inappropriate infrastructure and encourages the erosion of traditional social and cultural ties. This is the case in Bali.

> No more than a dozen surfers had chanced the trip from Kuta that morning so the line-up [at Uluwatu; Figure 11.1] was eerily deserted. The sun, even at 10 a.m., was searing through my white rashguard like a laser and the glare off the green steamy water made little fluorescent worms of light dance in my periphery. Surrealistic glowing waves looped over and spun down the reef at an elegant pace. It was a scene of such intense phosphorescent beauty as to approach a hallucination. After a short paddle up the line I found a likely take-off zone near a small pod of Brazilian surfers. I waited for an auspicious wave with a foreseeable end to stand up, and stroked into my dream.
>
> (Barilotti, 2002: 92)

> High atop the limestone temple steps, I gazed down at the long Indonesian lines feathering 500 feet below. The balmy aromatic smell of sandalwood incense wafted up from an unseen worshiper and the peevish little temple monkeys [Figure 11.2] chatted me up for a handout.
>
> (Barilotti, 2002: 93)

Attracted by the surfing environments and culture of Southeast Asia, I took a trip from the UK to Bali and its surrounding islands. Despite delivering a course on sustainability in my home institution, the direct environmental consequences of my flight to Asia failed to register during the anticipation phase of my adventure. The promise of 'phosphorescent beauty' drowned out the more abstract and

Figure 11.1 Uluwatu, Bali, Indonesia.
Source: author's image.

external repercussions of my dream trip. However, despite these consequences being out of sight and out of mind, they nevertheless were considerable. Using the Centre for Alternative Technology's[1] Carbon Gym software, a calculator which estimates the tonnage of carbon produced through an individual's lifestyle choices, it was possible to approximate the extent and significance of my travel to the waves of Bali (see Table 11.1).

As Table 11.1 illustrates, the Carbon Gym software ascribed an annual tonnage of carbon to key aspects of my daily lifestyle. Heating, cooking, food, shopping and regular transport are included in these activities, alongside commercial jet travel. My own personal carbon contribution totalled 15.11 tonnes per annum through this exercise (this compares to a UK average estimated at 10 tonnes per person per annum; a US citizen's average at 20 tonnes p.p.p.a.; and a Tanzanian average at 0.1 tonnes p.p.p.a.). If everyone on the planet contributed emissions to the atmosphere at the same rate as me, *six planets* equivalent to Earth would be required to absorb the carbon pollutants (see World Wildlife Fund, no date).

As Table 11.1 illustrates, however, the key contributor to my carbon emissions was commercial air travel. The ten tonnes figure not only contributes 66 per cent of my overall carbon emissions for the year, but is generated by my *one*

Figure 11.2 Temples of surf.
Source: author's image.

return flight from Heathrow, London, to Bali, Indonesia (via Singapore). This flight's carbon cost is over six times the per capita carbon footprint target for 2050 (1.5 tonnes p.p.p.a). Clearly, this flight to experience the surf cultures of Bali and its islands had significant environmental consequences, even before I set out paddling for a wave.

It is worthwhile noting at this stage the carbon cost of other flights, all estimated from the UK. A return flight to continental Europe (perhaps a weekender to Biarritz), would emit approximately 0.8 tonnes of carbon according to the Carbon Gym, and a trip to the US (perhaps a week in Santa Cruz) would emit four tonnes. Although one can debate the utility of carbon calculator software (centring on the use of carbon as a blunt proxy for sustainability in the round, the possibility for inconsistent metrics between different software and, as we will discuss later, the practices of offsetting that are implicitly encouraged through their use, see Gao *et al.*, 2013), it nevertheless throws into stark relief the external cost of the search for surfing utopia.

In my search for a place in the tube, I was also tempted to overlook other external consequences of my presence in foreign waters. The increase in popularity of surfing for both travellers and local populations is growing, and inevitably leads to overcrowding on both world-class waves such as Uluwatu, but also

Table 11.1 Carbon costs of a lifestyle, according the Centre for Alternative Technology's Carbon Gym software

Activity	Tonnes	UK average
Heating House	0.54	1.5
Water Heating	0.27	0.4
Cooking	0.12	0.3
Electrical Appliances	0.15	0.3
Food	2.11	2.0
Shopping	0.97	1.8
Waste	0.05	0.9
Transport	0.90	1.7
Air Flights	10.00	1.0

on other breaks throughout Bali. As I drew up to Padang Padang, the 'welcome' signs reminded me of the heightened competition existing between locals and tourists as they attempt to access surfing heaven (see Figure 11.3). Such competition is outlined in the following cases:

> It was such a perfect day to surf with the little one, so Wayan, a veteran surfer from Kuta, brought his 10-year-old son to Halfway Beach.... As the day wore on, groups of surf school students started crowding the water.... One of the students was already cruising fast toward his son and the two collided and were thrown into the water.... After checking on his son, Wayan, angry at what had happened, lunged at the instructor and a fight broke out between them.... News about the incident spread quickly, and stories of similar incidents followed.
>
> (Eric, 2011a: 20)

> The severe crowding in the line-up [referring specifically to The Point at Sorake Beach, Nias, but also experienced at Padang Padang] makes for a testy scene. Localism among the local surf kids has reared its myopic little head. Blatant drop-ins have become rampant and often backed up with insidious death threats.
>
> (Barilotti, 2002: 93)

Despite Fordham (2008) claiming that no one can ever own a wave, the rise in surfing's popularity, and the increasing numbers of global surf travellers seeking their own engagement with the place of foreign tubes means there is increasing competition over who can catch them. The utopia of finding empty waves waiting to be ridden is remote as surfers from around the globe congregate at culturally renowned hotspots. Whilst competition amongst surf tourists may be respectful (see Barilotti and the Brazilian pod above), competing with surfers from different locations and with different skill sets often results in frustration and animated confrontation (see Anderson, 2014b). Of all the external consequences produced by

Figure 11.3 Do I belong here? Padang Padang.
Source: author's photo.

being a global surf traveller, overcrowding is the most immediate to surfers; it directly affects their ability to commune with waves, and ultimately enjoy their stoke. Despite this it remains an open question whether an individual's quest for a place in the tube prompts reflection as to the causes and consequences of this over-crowding problem, and the personal role played in both these issues.

However, overcrowding is not the only negative consequence for local com-munities. The surfing gold rush inevitably attracts those seeking to cater for and exploit it. As a result, a second, economic, wave of incomers is generated as entrepreneurs, developers and businesses seek to extract dollars from surfing tourists. In many cases, the extent and nature of this second wave often washes away the local cultures existing in places like Bali. Regions such as Kuta and Legian have become notorious for their 'rampant development, low brow night-life and crass commercialism' (Lonely Planet, 2010: 269); although offering a 'tropical paradise' on the waves, these regions have become a prime example of 'the destructive effects of tourism' on the shore (Lonely Planet, 2001: 234). Bar-ilotti sums up the development of Kuta:

> Kuta Beach started out as a drowsy little fishing village in the 1930s, cater-ing to a small number of vacationing European colonialists. Its surf potential was discovered by Australians in the mid-1960s. Since then, it has morphed into a fully-fledged surf ghetto on a par with Huntington Beach or the North

Shore.... In our blind zeal to set up insular surf enclaves, we parachute advanced technologies into third-world economies and set up brittle unsustainable infrastructures. The list of soiled third-world surf paradises [like Kuta] ... is long and growing.

(2002: 92)

On nearby Gili Trewangen (Figure 11.4), an island off the northwest coast of Lombok and two hours by boat from Bali, the worst excesses of tourist development have been avoided to date. Gili Trewangen has both surfing breaks and excellent scuba diving opportunities (see Figure 11.5). Yet here too, the influx of global surf travellers has led to an increasing strain on ecosystemic resources. As Gili Eco Trust state,

Following the same pattern as in many other regions in the world, the pollution generated by [surf] tourists and the local population ... have worsened the critical situation already generated by several other factors of degradation of the environment such as the global warming, violent storms, and deforestation.

(Gili Eco Trust, no date)

Figure 11.4 Surf shacks on Gili Trewangen.
Source: author's photo.

Figure 11.5 Water world resources on Gili Islands.
Source: author's photo.

An island that has limited, and diminishing, freshwater supplies caters for surf tourists who often expect the relative luxury of freshwater showers, alongside Western foodstuffs and beverages. On islands without First-World infrastructure, such expectations lead to all resources being shipped in from the mainland (see Figures 11.6 and 11.7), and waste materials being shipped back again (Figures 11.8 and 11.9).

The unsustainability of these luxury items and their transport mirror the movement of the global surf travellers themselves; mobile due to privilege, preference and aspiration, rather than absolute need. The movement of people and foodstuffs creates negative environmental consequences that the holidaying users are often ignorant of. This, as Ponting *et al.* state, is a model of life that is 'market focused, economically neo-liberal and disconnected from the local place and people' (2005: 145). In practice, this model undercuts the apparent utopia existing on islands such as Gili Trewangen. It necessitates polite reminders to tourists from local populations about the negative consequences of their Western cultural preferences (see Figure 11.10), and attempts to define how a responsible surf traveller should behave when at a destination (see Figure 11.11).

Allied to specific environmental costs, it is also possible to witness the signs of cultural colonialism as a consequence of the presence of surfers in

Figure 11.6 Shipping in commodities.
Source: author's photo.

Figure 11.7 Bringing on water.
Source: author's photo.

Figure 11.8 Waste.
Source: author's photo.

Figure 11.9 Shipping out waste.
Source: author's photo.

communities such as Gili Trewangen. All travellers are vehicles of change, bringing with them cultural attitudes, expectations and commodities that can enhance, but also erode the cultural traditions of host societies. This may occur in mundane ways, such as the spread of exotic clothing, alien foodstuffs or language (see Figures 11.12 and 11.13), or the rise in social activities that would be traditionally disciplined and (b)ordered (see Anderson, 2015; Figures 11.14 and 11.15).

Across the archipelagos of Southeast Asia, including islands such as Nias and Bali, the creeping cultural colonialism that is present in Gili Trewangen is replicated, and is often more intense. Summarised by Barilotti, across the region, '4000 years of ancient animistic squat culture [has] now smacked straight into Western techheavy materialism' (2002: 92). This has led to the cultural influence of Western society colonising the traditions of regional and local cultures, as Barilotti explains with respect to Nias:

> the effect of surf tourism on the Niah, a proud, warlike tribe once notorious for their headhunting and elaborate costumed rituals, has sped the erosion and disappearance of traditional ways. Twenty-five years of cashed-up westerners tramping through Lagundi village has seduced the local youth with lurid Baywatch fantasies of the North American high life.
>
> (2002: 93)

Figure 11.10 Respect environment.

Source: author's photo.

Figure 11.11 How to be a responsible tourist.
Source: author's photo.

In Bali, as local journalist Eric reports, 'for centuries, Balinese women have obe-
diently carried the responsibilities assigned by tradition. Now, however, tradition
is becoming increasingly compatible with modern life. Women are beginning to
ask questions about their own destinies' (2011b: 30). Whilst Western ideas of
feminism may be seen as progressive when compared to Southeast Asian patri-
archies, the imposition of foreign ideas involves cultural imperialism and coloni-
sation. Substituting long-held traditions for industrialised poverty or employment
in hawking, prostitution and casual labour, all in the service of surf tourism,
appears to be a dubious advancement.

Sustainability, the individual surfer and beyond

We have seen, therefore, that beyond the internal consequences of surfing – the
stoke gained from the place of the tube and the wanderlust satiated through
global surf travel – a number of external consequences are also generated that
undermine the utopian spaces of surfing. In general it appears that for many
global surf travellers issues of sustainability are not a matter of priority.

Figure 11.12 Respect culture.

Environmental, social and cultural costs of surfing activity do not register highly for travellers, and the notion of sustainability, reflecting the situation in the broader population, is not a fundamental concern. Thus whilst sustainability as a moral, social and environmental goal is necessary, it is a language that doesn't 'speak to surfers'; it doesn't directly affect their internal goal – the experience of stoke, and their 'own sense of self' is not tied to the 'stability' or otherwise of their shore side destination (Tuan, 2004: 48). Sustainability asks surfers to respond to a morality that appears to have no place in the waterscape of their lives. Reflecting the 'market focused, economically neo-liberal' model of industrial society, they have externalised the negative consequences of their activity, effectively isolating their individual self from the broader societal, geographical and community ties that define it.

Nevertheless, the individualism of surfing may not be so antithetical to the principles of sustainability. Sustainability is increasingly framed in terms of the individual, and the isolated citizen is now often seen as a key contributor to the environmental crisis and any sustainable future (Burkitt and Ashton, 1996; Eden, 1993; James and Scerri, 2006; Scerri, 2009). Connections have been made between individual decisions to drive and fly, burn fossil fuels at varying rates, invest in certain developments or consume particular commodities, and the negative consequences these decisions have for the state of the environment. As Spaargaren and Mol put it:

Figure 11.13 Westernising culture.

> The individual is no longer seen as 'the small polluter' whose contribution
> can be dismissed in the light of the huge impacts of big industrial polluters
> ... the relative contribution of [individuals has now] come to the fore as the
> main obstacle for realising ambitious environmental goals.
>
> (2008: 354)

In this light, the individual(ist) preoccupations of surfers could be harnessed to
drive sustainability. If environmental destruction, global warming and sea level
rise were clearly understood as direct threats to the existence of certain breaks
and waves, then becoming part of a movement that adopts more sustainable
practices may make sense to the hedonistic surfer. Self-interest for the place of
the tube and the practice of travel to get there could drive more environmentally
and socially just practices.

Yet not all surfers are pathological. Indeed through their own practices many
become aware of the connections, commitments and responsibilities that can
develop between humans and their geographical territories (be they littoral or
terrestrial in the conventional sense). For many surfers, the activity is not simply
about skill development, or adrenalin shot, but also about the relational sensibility gained from communing with waves and feeling part of this elemental convergence. Whilst some articulate this experience in quasi-religious terms (Taylor,
2007a, b, c, 2008; Anderson, 2013), this geographical connection begins to

Figure 11.14 Drug culture.

Source: author's photo.

Note
The language chosen for these signs, and the expectation that travellers will not be conversant in Bahasa Indonesian.

Figure 11.15 Alcohol.

motivate a will to protect local areas, and support campaigns to conserve maritime resources. Here surfers are not isolating themselves from their broader social and environmental context, but rather redrawing the boundaries of the self; extending them to incorporate the various shores, waves and people that assemble together to enable their surfing rituals.

Moving beyond the individualist surfer thus requires that the self-centred (perhaps even selfish) participant is transformed into a citizen that extends their sense of self to include local cultures and environments. This extension resonates with the self-realisation doctrine of transpersonal ecology (see Fox, 1995). This posits that the boundaries between people and place are flexible, and that humans improve their well-being when they adopt a self-concept that includes friends, family, other humans, their contexts of everyday life and even the broader non-human environment. This model of the 'extended self' is currently most identifiable within surfing culture in relation to surfer's home breaks, with many surfers sensing that 'there is a close link between physical geography, stories and bodies in the act of surfing' (Evers, 2010: 47). As Evers goes on to outline:

> Each watery [encounter tied me] more tightly to the local environment. It became more than a place; it became my turf. The boundary between the place and me blurred; in fact the boundary was erased.
>
> (2010: 77)

For sustainability to be realised within surfing culture, this sense of an extended self that blurs with the broader physical and cultural environment needs to be adopted not simply in relation to a surfer's home break, but also to the place of the tube abroad. If surf culture is serious about minimising its effects on the planet and its people that come together to form the experience that we love, surf culture must be prepared not only to realise the benefits, but also to absorb the costs, of changing our personal lives in line with sustainable ideals. Through knowing the harm that certain activities cause to the environment we must be prepared to 'rearrange our wanting' (after Wilding, 2003) as discrete individuals, and align our lifestyles in line with what is best for the extended self. In this sense, this process would be less a matter of depriving ourselves of 'goods' but, due to the re-formation of our sense of identity, more a question of extended 'self respect' (see Wilding, 2003; also Fox, 1995; Naess, 1989).

Part of this 'rearranging' of surf culture could involve the reflection on the appropriateness of global surf travel actioned by jet-engine aircraft. Due to severely damaging consequences of commercial jet travel, in theory at least surf culture should cease to encourage global surf travel premised on this form of transport. The absolute removal of commercial jet travel from the rituals of surfing would be a radical change to the cultural orthodoxy; it would mean that surfers like Ouhilal would be unable to access their utopic Groundhog days at short notice. Yet overland trips or slow boats are not precluded from this solution, and such travel may offer new forms of adventure and new surf spots for the intrepid surf traveller. Indeed, perhaps it is not such a big leap to begin thinking more ecologically in respect to the questions of whether to travel to, and surf in, a particular place. Why should this question be out of bounds to surf culture? To echo common rhetoric often used in relation to environmental issues (but often attributed to John F. Kennedy (see Popik, 2010), if we do not think about it, then who? And if not now, then when? (see also Wilson, 2015).

Pipedreams? Surfers as pragmatic coyotes

In all likelihood, however, such radical alternatives are unlikely to gain immediate approval within surf culture, despite its ecological necessity. More likely would be the adoption of a 'politics of pragmatism' in relation to surfing lifestyles (Anderson, 2012b). A politics of pragmatism in relation to surfing lifestyles involves individuals adopting a pragmatic strategy of 'incremental integration', aligning their extended self in relation to the greater good which now includes the internal *and* external consequences of their surfing rituals, at both the local and global scale. In this strategy, individuals choose to consistently reflect on the consequences of their actions, and commit to adopt those changes that will both enhance sustainability and are realistically possible. In terms of the practice of travel, this may mean that the number of overseas trips by commercial air travel is reduced, but not necessarily eradicated completely. For those trips that remain, carbon offsetting could be considered. Carbon offsetting involves making financial payment to tree growing projects that helps to absorb the carbon emitted from flights, or renewable energy schemes and technological developments that offer longer-term solutions to existing unsustainable practices. In terms of the place of the wave, such a strategy would also extend a surfer's 'self-respect' to include local peoples and their culture. In practice this may mean that surfers do not expect First-World luxury to be available, necessary or desirable in all facilities and destinations; that locals have a provenance on a wave and deserve respect and sometimes deference; that sensitivity is required to the ways in which surf culture can accelerate the erosion of indigenous ways of life; and that although different to that of the global surf traveller, local traditions are not necessarily inferior or warrant eradication at the hands of an oblivious or negligent tourist. Respect on the waves, in short, should be extended to respect for the environment and culture on the shore.

Such a pragmatic politics resonates strongly with the notion of coyote environmentalism (see Anderson, 2010). As a prefix for any subject position the 'coyote' was introduced to academia by Haraway (1988). It refers to any identity encultured with tricky, subversive and problematic capacities. Haraway borrows this metaphor from Native American culture where it symbolises someone who is 'quick, sly, [and] lecherous, [but] also a carrier of valuable knowledge, a rebel, a survivor' (Snyder 2007: 45). This 'trickster' image may be attractive to the countercultural, maverick, and adventurer caricature that is popular within surf culture, but it is also useful for summarising the contradictions and multiplicities inherent within adopting sustainability at the individual level. The coyote is a subject position that embodies conflicts, problems, solutions and possibilities. As Snyder (2007) suggests with reference to intelligence and morality, the position of the coyote, 'reminds us that wisdom and foolishness are often mingled hopelessly together like ghee stirred into milk' (2007: 45). Haraway (1988) suggests that the coyote can stand for an individual who interrogates their positionings, is happy to be accountable, and can 'construct and join rational conversations and fantastic imaginings that change history' (1988: 586). Such a position may enable surfing and sustainability to be more than simply a pipedream.

Conclusion

This chapter has argued that an overwhelming focus on the internal and individual consequences of surfing has led to a marginalisation of its external consequences. Without appropriately addressing these often negative externalities, surfing utopia will always be shaded by social and environmental consequences. As a result of ignoring these external consequences of our surfing rituals, surfing utopia is a place that cannot truly be (see Figure 11.16).

However, by transforming the surfer into an individual who jettisons an individualised, self-centred and -obsessed identity, for an extended self that includes interest in and responsibility for the broader communities and environments that support their water worlds, it may be possible for the external consequences of surfing to become integrated more appropriately within surf culture. From this beginning, a pragmatic, coyote, approach to greening surf culture could be

Figure 11.16 Utopia: a place that cannot be?

adopted, where surfers attempt to minimise the negative consequences of their actions whilst still committing to a life in the tube. In this way, surfers may be able to extend their commitment to the transience of stoke, reorienting their 'liquid lives' (see Bauman, 2000) around the cultures and places that make this relational sensibility possible. As a result, a 'moral conversation' is opened up (Hobson, 2001) that can generate a new model of 'good' surfing which extends the surfing code and self to include the broader concerns of sustainability, and translates it from these pages and into practice. In this way it may be possible for surfers to generate a respect for and sustainable use of local places and their resources, in full recognition that these are as fragile and fleeting as the surf itself, and through their union produce a new sensibility which gives meaning to their lives.

Note

1 The Centre for Alternative Technology is based in Machynlleth, UK. It was established in 1973 and modelled on US hippy communes, but has become the 'leading eco-centre' in Europe (CAT, 2007a: no page), 'dedicated to eco-friendly principles' and functions as a ' "test bed" for new ideas and technologies' (CAT 2007b: no page). The Carbon Gym was one of these technologies, designed to record an individual's carbon footprint.

References

Anderson, J. 2010 From 'zombies' to 'coyotes': environmentalism where we are. *Environmental Politics.* 19 6 973–992.

Anderson, J. 2012a Relational places: the surfed wave as assemblage and convergence. *Environment & Planning D: Society & Space.* 30 4 570–587.

Anderson, J. 2012b Managing trade-offs in 'Ecotopia': Becoming green at the 'Centre for Alternative Technology. *Transactions of the Institute of British Geographers.* 37 2 212–225.

Anderson, J. 2013 Cathedrals of the surf zone: regulating access to a space of spirituality. *Social and Cultural Geographies.* 14 8 954–972.

Anderson, J. 2014a Exploring the space between words and meaning: understanding the relational sensibility of surf spaces. *Emotion, Space & Society.* 10 27–34.

Anderson, J. 2014b Surfing between the local and trans-local: identifying spatial divisions in surfing practice. *Transactions of the Institute of British Geographers.* 39 2 237–249.

Anderson, J. 2015 *Understanding Cultural Geography: Places and Traces.* Second Edition. Routledge: London and New York.

Anderson, J. and Peters, K. eds. 2014 *Water Worlds: Human Geographies of the Ocean.* Ashgate: Farnham.

Barbieri, C. and Sotomayor, S. 2013 Surf travel behavior and destination preferences: an application of the Serious Leisure Inventory and Measure. *Tourism Management.* 35 111–121.

Barilotti, S. 2002 Lost horizons: surfer colonialism in the twenty-first century. *The Surfers' Path* 33 30–39.

Bauman, Z. 2000 *Liquid Modernity.* Polity Press: London.

Buckley, R. 2002 Surf tourism and sustainable development in Indo-Pacific Islands. 1. The industry and the islands. *Journal of Sustainable Tourism,* 10 5 405–424.

Burke, E. 1776 *Philosophical Enquiry into the Origin of Our Ideas of the Sublime and Beautiful.* Dodsley: London.

Burkitt, B. Ashton, F. 1996 The birth of the stakeholder society. *Critical Social Policy.* 16 3–16.

Casey, S. 2011 *The Wave.* Anchor Books: New York.

Centre of Alternative Technology (CAT). 2007a *What Do We Do?* www.cat.org.uk/information/aboutcatx.tmpl?init=1 Accessed July 2007.

Centre of Alternative Technology (CAT). 2007b *How CAT Started.* www.cat.org.uk/information/aboutcatx.tmpl?init=4, a permanent visitor centre. Accessed July 2007.

Colas, A. 2001 *The World Stormrider Guide.* Low Pressure: Bude, Cornwall.

Csíkszentmihályi M. 1990 *Flow: The Psychology of Optimal Experience.* Harper and Row: New York.

Darnton, A. 2004 *The Impact of Sustainable Development on Public Behaviour* [online]. *Report 1 of Desk Research Commissioned by COI on behalf of DEFRA.* www.sustainable-development.gov.uk/documents/publications/deskresearch1.pdf Accessed May 2004.

Doherty, S. 2007 *The Pilgrimage.* Penguin Books: London.

Eden, S. 1993 Individual environmental responsibility and its role in public environmentalism. *Environment and Planning A.* 25 1743–1758.

Eric 2011a Turn of the tide. *The Mag.* 37 May 20–22.

Eric 2011b Women of Bali. *The Mag.* 37 May 29–32.

Evers, C. 2010 *Notes to a Young Surfer.* University of Melbourne Press: Melbourne.

Ford, N. and Brown, D. 2006 *Surfing and Social Theory: Experience, Embodiment and Narrative of the Dream Glide*. Routledge: London.

Fordham, M. 2008 *The Book of Surfing: The Killer Guide*. Bantam Press: London.

Fox, W. 1995 *Towards a Transpersonal Ecology: Developing New Foundations for Environmentalism*. Resurgence: Totnes.

Gao, T. Liu, Q. and Wang, J. 2013 A comparative study of carbon footprint and assessment standards. *International Journal of Low-Carbon Technologies*. doi: 10.1093/ijlct/ctt041.

Gili Eco Trust. no date Gili Eco Trust News. www.giliecotrust.com Accessed May 2015.

Graber, L. 1976 *Wilderness as Sacred Space*. Association of American Geographers Monograph Series: Washington, DC.

Haraway, D. 1988 Situated knowledges: the science question in feminism and the privilege of partial perspective. *Feminist Studies*. 14 3 575–599.

Hobson, K. 2001 Sustainable lifestyles: rethinking barriers and behavioural change. In Cohen, M. and Murphy, J. eds *Exploring Sustainable Consumption: Environmental Policy and the Social Sciences*. Pergamon: London. 191–209.

Ingold, T. 2007 *Lines: A Brief History*. Routledge: Abingdon, UK.

James, P. and Scerri, A. 2006 Globalizing life-worlds: the uneven structures of globalizing capitalism. In Darby, P. ed. *Postcolonizing the International*. University of Hawai'i Press, Honolulu, HI. 73–102.

Lonely Planet. 2001 *South East Asia on a Shoestring*. Lonely Planet: London.

Lonely Planet. 2010 *Indonesia*. Lonely Planet: London.

Macnaughten, P. and Jacobs, M. 1997 Public identification with sustainable development: investigating cultural barriers to participation. *Global Environmental Change – Human and Policy Dimensions*. 7 1 5–24.

Martin, S. A. and Assenov, I. 2012 The genesis of a new body of sport tourism literature: a systematic review of surf tourism research (1997–2011). *Journal of Sport and Tourism*. 17 4 257–287.

Naess, A. 1989 *Ecology, Community, Lifestyle*. Cambridge University Press: Cambridge, UK.

Otto, R. 1970 *Mysticism East and West: A Comparative Analysis of the Nature of Mysticism*. Macmillan: New York.

Ouhilal, Y. 2011 Utopia: a place that cannot be. *Carve* 123 May 88–97.

Ponting, J. 2008 Consuming Nirvana: an exploration of surfing tourist space (Doctoral dissertation). University of Technology, Sydney.

Ponting, J. 2014 Research notes: comparing modes of surf tourism delivery in the Maldives. *Annals of Tourism Research*. 46 163–165.

Ponting, J. and McDonald, M. 2013 Performance, agency, and change in surfing tourist space. *Annals of Tourism Research*. 43 415–434.

Ponting, J. and O'Brien, D. 2015 Regulating 'Nirvana': sustainable surf tourism in a climate of increasing regulation. *Sport Management Review*. 18 99–110.

Ponting, J. McDonald, M. and Wearing, S. 2005 Deconstructing wonderland: surfing tourism in the Mentawai Islands, Indonesia. *Society & Leisure*. 28 1 141–162.

Popik, B. 2010 'If not us, who? If not now, when? April 25. www.barrypopik.com/index.php/new_york_city/entry/if_not_us_who_if_not_now_when Accessed May 2015.

Scerri, A. 2009 Paradoxes of increased individuation and public awareness of environmental issues. *Environmental Politics*. 18 4 467–485.

Seghezzo, L. 2009 The five dimensions of sustainability. *Environmental Politics*. 18 4 539–556.

Snyder, G. 2007 *Back on the Fire: Essays*. Shoemaker & Hoard: Emeryville, CA.

Spaargaren, G. and Mol, A. 2008 Greening global consumption: redefining politics and authority. *Global Environmental Change*. 18 350–359.

Stranger, M. 1999 The aesthetics of risk: a study of surfing. *International Review for the Sociology of Sport*. 34 3 265–276.

Taylor, B. 2007a Focus introduction: aquatic nature religion. *Journal of the American Academy of Religion*. 75 4 863–874.

Taylor, B. 2007b Surfing into spirituality and a new, aquatic nature religion. *Journal of the American Academy of Religion*. 75 4 923–951.

Taylor, B. 2007c The new aquatic nature religion. *Drift*. 1 3 14–23.

Taylor, B. 2008 Sea spirituality, surfing, and aquatic nature religion. In Shaw, S. and Francis, A. eds *Deep Blue: Critical Reflections on Nature, Religion and Water*. Equinox: London. 213–233.

Tuan, Y. F. 2004 Sense of place: its relationship to self and time. In Mels, T. ed. *Reanimating Places*. Ashgate: Aldershot. 45–56.

Warshaw, M. (ed.) 2004 *Zero Break: An Illustrated Collection of Surf Writing 1777–2004*. Harcourt: Orlando, FL.

West, P. 2014 'Such a site for play: this edge': surfing, tourism, and modernist fantasy in Papua New Guinea. *The Contemporary Pacific*. 26 2 411–432.

Wheaton, B. 2005 Selling out? The commercialisation and globalisation of lifestyle sport. In Allison, L. ed. *The Global Politics of Sport*. Routledge: London. 140–162.

Wilding, J. 2003 *Dinner, the Planet and Human Rights*. www.wildfirejo.org.uk/feature/display/21/index.php Accessed July 2008.

Wilson, J. 2015 Transitions to sustainability: 'If not us, then who?'. *Sustainable Stoke: Transitions to Sustainability in the Surfing World*. University of Plymouth Press: Plymouth, UK. 50–54.

World Commission on Environment and Development. 1987 *Our Common Future*. Oxford University Press: Oxford.

World Wildlife Fund. No date *Carbon Footprint Calculators*. http://footprint.wwf.org.uk/ Date Accessed May 2015.

12 Culture, meaning and sustainability in surfing

Neil Lazarow and Rebecca Olive

Introduction

Surfing is at a(nother) crossroad. In the past ten years there has been an explosion of thought and literature on the evolution of culture, meaning and sustainability in surfing. More women are surfing than ever before, there is a greater variety (and acceptance of) surfcraft than in any decade previously, urban water quality is vastly improved in some of surfing's traditional bastions, and there has been an evolution in the forms and manner in which surf media are generated and consumed. There are also more surfers contesting for waves than at any other time in human history, greater pressure on many coastal resources and an expansion of surfing's carbon and cultural footprint across the planet; and there is no sign that this is slowing down. In this chapter, we take a moment to reflect on these changes; to consider the good, the bad and the ugly, and to ask whether 'the surfing community', if that even exists in any meaningful form today, is in good health.

Until recently there was relatively little serious introspection about culture, meaning and sustainability in surfing (Lazarow, 2010). These analyses have not only come from researchers. In an editorial for *Kurungabaa: A Journal of Literature, History and Ideas from the Sea*, former *Tracks* magazine editor Tim Baker (2008) argued that surf magazines were beholden to advertisers, limiting the independence of the surf media to engage in credible discussions around issues such as environment, surf colonialism, gender, design and the role of industry. Despite a connection to issues of ecology and sustainability in 1970s surf media – such as the notorious Australian magazine *Tracks* and the iconic film *Morning of the Earth* – contemporary surf media's exposure of these issues in the 1980s, 1990s and 2000s was limited, largely expressed through 'green issues' or feature stories on endangered or threatened waves, reflecting the focus and attention that surfing conservation organisations such as Surfrider Foundation (est. 1984) and Surfers Against Sewage provided. Exceptions to this are niche magazines such as *The Surfer's Path* (1997–2013), *Kurungabaa* (2008–2013), the Groundswell Society's annual publication and *The Surfers Journal* (to a lesser extent), which engaged with issues of ecology and environmental activism.

Since the early 2000s a number of scholars have argued that to think about surfing means that one must explore the symbiosis of people and place (Anderson, 2009, 2014; Evers, 2008a, 2008b; Lazarow, 2010; Olive, 2015; Satchell, 2007, 2008). As Lazarow points out, without people, waves remain 'ecosystem functions', a part of the natural order that aligns and realigns coastlines around the world (Lazarow, 2010). However, by including the understandings of, and relationships between surfers, waves become assets around which communities, economies, cultures, experiences and ecologies are born and lived, something to admire, protect, cherish, fear, prize and negotiate. That is, surf breaks are geographical, historical, social and cultural, and need to be understood as linked to community, health, economics, politics and identity.

Empirical information

The empirical information in this discussion comes from research conducted by Neil (Lazarow) from 2006 to 2008. In this project, a series of online and face-to-face surveys were administered, with over 800 participants responding. The results discussed in this chapter are drawn from an international sample of 470 respondents for which Australian (approximately 43 per cent) and US (approximately 40 per cent) participants constitute the majority of respondents (i.e. the web-based survey), and a related study on Australia's Gold Coast, which consisted of 480 returned surveys, half of which were face-to-face and half online.

Data was collected through a combination of face-to-face surveys and a web-based survey instrument. The bespoke face-to-face survey was administered on the Gold Coast beaches in Queensland, Australia by a number of trained research assistants at pre-determined locations in order to achieve good coverage across surf beaches. The survey website (goodfortheplanet.com), which carried a slightly more general survey, was advertised via electronic media release picked up and redistributed through a number of surf forecast/surf media websites, network emails and email dispatches through major surfing NGOs, through other websites (for example, www.coastalwatch.com), word of mouth, through existing networks and through the distribution of advertising flyers at major events such as the Quiksilver Pro surfing competition.

Participants taking part in the survey through the web-based instrument 'self-selected' by responding to one of the multiple electronic invitations to participate in the survey. The process of 'self-selection', whereby individuals are made aware of and then choose to fill in the online survey, may generate a result that is biased towards those individuals who participate in surfing more frequently. With approximately 75 per cent of all surveys reported in this study using the online survey instrument, there is a risk that the results may not present a balanced view of surfing activity, and may instead focus on the views of 'core' participants. In this way, the survey data can be read to most represent the views of experienced surfers who surf regularly and consider themselves committed participants. Focus on regular surfers is a common feature of research about surfing (see for example Olive, 2015; Stranger, 2011). In this study the use of both face-to-face and online surveys was useful as an effort to reduce the risk of over-representation by this group (Lazarow, 2010).

The Gold Coast survey instrument had five question sections: demography, surfing effort, expenditure, motivation, personal information. The global online survey also contained five sections: demography, personal information, surfing effort, motivation, surfing culture today.

Approach

In this chapter, Neil's data is critically developed in collaboration with Rebecca, whose research adopted a feminist cultural studies approach to cultural gender power relations in surfing culture in the surf and on social media (Olive, 2013, 2015; Olive *et al.*, 2015). In her research about women and surfing, Rebecca applies understandings of culture, power, ethics and pedagogy operating through embodied subjectivities to explore processes of knowledge production and change. In this collaboration, we negotiate our different disciplines (public policy, political science and cultural studies), methodological approaches (quant-itative and qualitative ethnographic approaches) and paradigms (positivist, con-structionist and post-structural feminist), as well as shared interests in environmental and social justice, to think through the same empirical informa-tion. Drawing on this data, our ambition in this chapter is to take a closer look at key aspects of the contemporary surfing world – culture (lived experience and media), meaning, place and sustainability. By exploring the intersections between these, we aim to think about sustainability in more holistic terms, as relating to surfing culture, localism, relationships (to place and people) and coastal / surf break management. In developing this discussion we also draw on our research, lived experience and stoke for surfing and the beach.

Our discussion contributes to the growing number of voices critiquing estab-lished understandings of surfing history and hierarchies, as well as the powerful role traditional media and industry have had in developing surfing's cultural iden-tity (Booth, 2012; lisahunter, 2017; Olive, 2015). These critiques have been important because surfing has moved rapidly past the point where it can be captured as a 'whole' in any one discussion. Of course, it has never been possible to list the many themes or faces of surfing. For example, Finnegan, in the foreword to *The Encyclopaedia of Surfing*, suggests that surfing as a subject really has no edges: it has 'no center, no institutional structure … and that even if organised competition and the loose set of enterprises known as the surf industry were to vanish tomorrow nearly all surfers would still carry on with their wave-riding lives' (in Warshaw, 2004, p. vii). We aim to engage with this loose-ness to explore how surfers understand surfing culture, as well as its limitations.

Serious leisure and attachment to place

Manning (1999, p. 222) contends that 'through a process of socialization, recrea-tionists may acquire specialist knowledge, skills, attitudes, and norms that define their progression from beginner to expert'. More specifically, Miller *et al.* (1999) suggest that in acknowledging the ritual potential of surfing, it is important to understand the motivation and passionate commitment that both locals and tourists exude in the pursuit of their favourite coastal activity. Such commitment is similar to what Stebbins describes as 'serious leisure', which is the noted intense levels of personal involvement and high levels of technical competence tantamount to professionalism. Stebbins notes that (1979, p. 13):

> Serious leisure can be defined as the systematic pursuit of an amateur, hobbyist, or volunteer activity that is sufficiently substantial and interesting for the participant to find a career there in the acquisition and expression of its special skills and knowledge.

Specialisation, the practical manifestation of serious leisure, argues Manning (1999, pp. 231–232, adapted from Little, 1976 and McIntyre and Pigram, 1992), can be measured in a number of different ways including behavioural (for example experience level, type of equipment used, preferred settings), psychological (for example involvement/commitment, centrality to lifestyle) and cognitive (for example knowledge, skill/expertise) measures. Capital – cultural, social and otherwise – argue Ford and Brown (2006), must be developed at some point through surfing competence; however, it is not necessary to maintain a 'hardcore' lifestyle in order to maintain capital. The path to specialisation and the relationship a person has with a particular activity as they develop from novice towards specialist speaks to the attainment of more than just high levels of competency, it also allows for cultural recognition by other users.

Surfers, like other lifestyle and action sports participants, express their individual personalities and being through their surfing, or 'leisure identity' (Kyle *et al.*, 2007, p. 12) as well as their beach-related behaviour and language. Manning (1999) suggests that more highly specialised recreationists will have more highly developed preferences with respect to both the type of environment and the kind of social interaction they prefer. As specialisation in an activity increases (along with age), surfers then, should be more likely to take an interest in the preservation of their leisure spaces as well as other aspects of their leisure experience. In surfing, Lazarow (2010) suggests that it is local residents or regular visitors who are more likely to become involved in on-ground campaigns, whereas occasional visitors or national industry organisations are less likely to be aware of or take an interest in changes at the local scale (Anderson, 2009, 2014; Olive, 2015; Satchell, 2008; Wheaton, 2007). In this way, again, non-residents or less committed surfers are unfairly seen as having less commitment to the preservation of the well-being of coastal communities and environments (Olive, 2015).

In becoming involved in collective expression or action, surfers create what Irwin (1977) has termed 'scenes', and what has more recently been located as style. This 'personal relevance' is critical to understanding the motivations of involved recreationists (Kyle *et al.*, 2007). This type of expression has been well documented in Australia, the USA and the UK (for example, Booth, 2001;

Boullon, 2001; Ford and Brown, 2006; Kampion, 1997; Ormrod, 2007; Young *et al.*, 1994), including through psychology and behaviour (Bennett, 2006; Young, 2000). Johnson and Orbach (1986) in particular present an interesting examination of the differences between surfers and fishers at Atlantic Beach, North Carolina and Newport Beach, California. The researchers found that where low-capital beach-based use took place, it was often the role of local cultural forces that determined the dominant activity. However, this was also influenced by commercial interests, public access, media and popular culture. Johnson and Orbach (1986) suggest that 'beach uses will become more homogenised as mass media and mass culture pervade local communities more and more'. Yet despite Johnson and Orbach's claims, beach and surfing scenes are far from one-dimensional, creating distinctions along the lines of board choice (Booth, 1995, 1999), sex/gender (Evers, 2006, 2008a; Olive, 2016), sexuality (Roy, 2013; Roy and Cauldwell, 2014) and race and ethnicity (McGloin, 2005; Preston-Whyte, 1999, 2002; Wheaton, 2013). That is, while easily reduced to physical cultures, these scenes have complex internal tensions and contestations that privilege some surfers over others. In response, recent films, books, magazines and online and social media have begun to present more diverse understandings and experiences of surfing. Preston-Whyte (2002) argues that for all surfers, surfing space is constructed around the idea of a surf break and the material environment must be included in an understanding of surfing. This is a critical point and the one on which the nature–society constructs around the visions for 'sustainability' must surely rest. Kyle *et al.* (2007, p. 118) argue that 'although setting plays a large role in determining what behaviours are possible within the spatial context, understanding the meanings people associate with a particular place provides insight on why individuals and groups value particular resources'. In other words, 'recreationists' themselves play a crucial role in the process of constructing experience. The 'consumptive orientation' of individuals within a particular user group (for example anglers or surfers) can vary markedly. There is no doubt that at the beginning of the twenty-first century, the coastal zone has become a highly contested space with many groups and individuals competing for access and preferential use rights to the sand and the nearshore zone and surfing can be impacted on in a number of ways, physically and environmentally (Challinor, 2003; Kelly, 1973) and culturally (Booth, 1995, 2001; Nazer, 2004; Preston-Whyte, 2002).

These tensions in surfing participation – the tendency to individualisation and consumption in lifestyle sports participants, and the embodied relationships and concern for the protection of place that participants regularly express – are an important issue in contemporary surfing. Ford and Brown (2006, p. 64) refer to surfing's mainstream 'coolness' as a form of 'subculture seduction', in which there is both extensive internal commentary as well as interaction with mainstream media. Understanding the multiple and sometimes competing values and motivations that drive these enthusiasts is important if surfers are to be further mobilised into supporting place-based and/or environmental causes. The responses to Lazarow's survey questions (discussed below) highlight these tensions. The rest of this discussion will explore two main sections of the survey – 'Meaning in surfing' and 'Social and environmental perceptions'.

Meaning in surfing

In the last major study of surfers' attitudes and preferences in Australia and New Zealand/Aotearoa, Kent Pearson (1979) interviewed 233 surfers (192 from Australia and 41 from New Zealand). Pearson's primary goal was to 'concentrate on the social and psychological "meanings" individuals find in their surfing activities' (Pearson, 1979, p. 70). Table 12.1 lists the top ten meanings or categories for involvement in surfing found by Pearson, for which the top three are: cathartic, fun and exercise.

Table 12.1 Top ten meanings or reasons for involvement in surfing activities

Category		*Total (n = 233)*	*Percentage*	*Example*
1	Cathartic, compensatory	83	35.6	Surfing is one way to get away from the artificiality of the establishment and appreciate real living
2	Fun, pleasure, joy, vivification	57	24.5	Surfing is having fun – letting it all hang out with mates
3	Exercise, health, fitness	54	23.2	Surfing is a clean, healthy, refreshing sport
4	Interaction with environment	50	21.5	Surfing is a way of appreciating nature by almost taking part in it
5	Geographical resources	42	18.0	The climate and availability of surf would be the two main factors why people surf
6	Vertigo, excitement, risk-taking	30	13.3	Surfing gives an excitement that comes from forms of speed and skill
7	Social experience – affiliation	29	12.4	Has a breed of 'guys and chicks' (not including posers), a great social life
8	Aesthetic experience	28	12.0	Surfing to me is an art – like music or poetry or painting – because you have your own style
9	Whole way of life	28	12.0	Surfing is a whole way of life.... It governs everything. Where you live, what you do with your spare time, who you mix with, your style of clothes
10	Technical skills challenges (physical prowess)	22	9.4	A collection of skills of almost unique application, providing numerous challenges

Source: Pearson (1979, pp. 71–76).

Motivation for surfing

In response to the survey question on motivation, approximately 830 respondents (n=800–831, depending on the question) answered the question on motivation for surfing. For this question, 56 per cent of respondents were Australian (70 per cent from the Gold Coast), 26 per cent from the US, 8 per cent from Chile, 5 per cent from NZ and the remaining 5 per cent from a range of other countries. Despite the large variances (distance apart, probable surf quality, demographics, etc.) it's clear that there are some catalysing themes and aspirations that surfers attach to the surfing experience.

The categories for motivation were developed by reviewing a range of national park user surveys, previous studies on surfing (Kelly, 1973; Pearson, 1979) and the author's personal experience. Insider knowledge is consistently located as offering important insights into participant-driven cultures like surfing, which rely on unspoken and embodied knowledges as a major form of knowledge transfer (Evers, 2006; Olive, 2016; Stranger, 2011). As such, an insider perspective offers much in terms of categorisation and analysis of data. As well as the eight categories listed in Figure 12.1, respondents were also given an opportunity to describe two other personal motivational forces (these are listed in Table 12.2). The results of the survey suggest that the top four categories (to be outdoors – 74 per cent, to relax – 73 per cent, to bond with nature – 58 per cent, fitness – 50 per cent) align closely with Pearson's findings.

Fitness (Pearson's third category), sport (Pearson's tenth category) and competition (rated by only 0.5 per cent of respondents to Pearson's survey – and last) were listed as separate categories in this survey. Interestingly, 51 per cent of respondents, the third highest response for all questions, stated that they

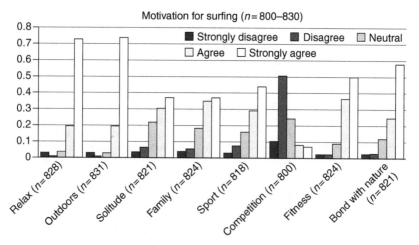

Multiple response options mean that values add up to more than 100%

Figure 12.1 Motivation for surfing.

Table 12.2 Motivation for surfing: open answers

Category	Total (n = 177)	Percentage
1 Cathartic, lifestyle, self-expression, motivation	50	28
2 Fun, enjoyment, mateship	39	22
3 Spiritual, religious	18	10
4 Technical, skill	15	8
5 Adrenalin	12	7
6 Bond with nature	9	5
7 Friends, meet people	9	5
8 Work	7	4
9 Family	5	3
10 Learn about the ocean	3	2
11 Sport, fitness	4	2
12 See women	3	2

disagreed that their motivation for surfing was 'competition'. When those who strongly disagreed with 'competition' being their motivation for surfing are added to this, the total response is 61 per cent. However, it is worth noting that 8 per cent and 7 per cent respectively (a total of 15 per cent) agreed or strongly agreed that one of their motivations for surfing was 'competition'. Allowing for some misinterpretation of the question by participants and despite a significant preference for non-competitive involvement registered by many respondents, it is apparent that more surfers are participating in competition nowadays than they were 35 years ago. The growth of professional surfing as well as club surfing in Australia, the USA and other countries seems to suggest that this is definitely the case. Government investment in a High Performance Surf Academy on the Gold Coast is also indicative of the success of the sports administrators to mainstream surfing.

For the two open categories section of this question, 177 answers were received (see Table 12.2). Given an opportunity for respondents to express themselves freely, the top two categories are the same as those identified by Pearson. Responses revolving around cathartic experiences (28 per cent) were highest; followed by responses related to fun, enjoyment and mateship (22 per cent); and then religious/spiritual experiences (10 per cent). If 'bonding with nature' and 'spiritual and religious' are combined then the total for this category is raised to 15 per cent of responses (see Table 12.2).

Surfers' social and environmental perceptions

More than half of those surveyed (54 per cent) in the global survey believe that the surf industry cares more about the environment today than it did ten years ago. At the country level, there was a significant difference between Australian and US respondents with 51 per cent of Australian respondents agreeing or strongly agreeing, compared to 74 per cent of USA respondents (see Figure 12.2). This substantial difference is indicative of the lack of homogeneity across the surfing population – even for seemingly similar outward surf cultures. Both Rebecca and Neil suggest that Australian surfing's anti-establishment culture that emerged and developed through the mid-late twentieth century still provides a strong (yet perhaps ironic since today this is arguably more of a marketing veneer than legitimate point of difference) backdrop against which surfers distinguish themselves in society. In the USA, the culture of corporate social responsibility amongst other related issues has enabled the surf industry to develop a far more symbiotic relationship with the 'recreational' surfing community in many locales – and this might be reflected in a more sympathetic view from survey respondents. Finally, and positively for Australia, issues such as water quality and beach access have been far more concerning for surfers in the USA, possibly offering greater opportunities for the surf industry to participate in campaigns and on-ground remediation activities.

Despite surfers' proximal relationship to nature, in this data it appears that many more surfers believe (35 per cent compared to 21 per cent) that as a whole surfers give less back to the planet than those involved in other (somewhat comparable) outdoor recreation user groups (see Figure 12.2). Participants' responses to the question on whether they'd get more stoke from teaching someone to stand up on a board for the first time or drop into a six foot wave, yielded some interesting differences across the Australian ($n=195$) and USA respondents ($n=182$). For example, 43 per cent of American respondents agreed or strongly agreed compared to 30 per cent of the Australians; and 25 per cent of American respondents disagreed or strongly disagreed, compared to 38 per cent of Australian respondents. On the surface, it appears that Australian surfers have a more individualistic approach to and sense of fulfilment from surfing.

In response to the question whether they were happy with the mainstream face of surf culture, about one-third of respondents strongly agreed or agreed, one-third were neutral and the remaining third either disagreed or strongly disagreed (see Figure 12.2). Overall, Australian respondents ($n=195$) were somewhat more positive (37 per cent) than US ($n=183$) respondents (25 per cent). This is an interesting result, and perhaps speaks to the fact that surfing as we know it is a broad church.

Recognising that a range of pressures on the coastal environment such as water quality and restricted access can affect the surfing experience and the health of surfers, two questions were designed to explore what types of issues might prevent a person from going surfing. The results are displayed in Figure 12.3 and Figure 12.4. Participants were provided with six scenarios (three biophysical and three social) and asked to rank each option on a scale of 1–6 from 'No way I'm going surfing' to 'Definitely going surfing'. For this question, 58 per cent of respondents were Australian (84 per cent from the Gold Coast), 29 per cent from the US, 6 per cent from NZ, and the remaining 7 per cent from a range of other countries.

Most likely to keep a person out of the water is the risk of contracting gastroenteritis, with 46 per cent of respondents indicating that they were unlikely (24 per cent) or definitely not (22 per cent) going surfing. The second factor most

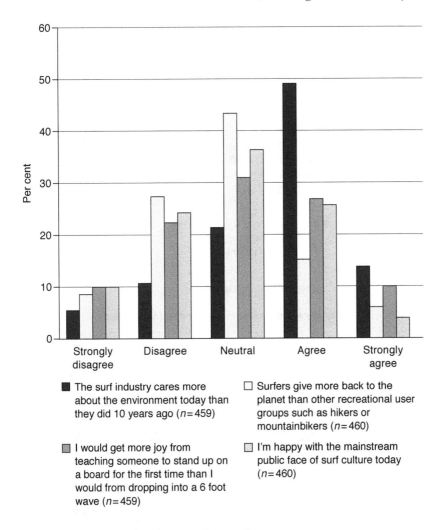

Figure 12.2 Surfers and environmental perceptions.

likely to keep a surfer out of the water is the likelihood of aggression in the line-up, with 44 per cent of respondents indicating that they were unlikely (15 per cent) or definitely not (29 per cent) going surfing. Third most likely to keep a surfer out of the water is the risk of contracting a skin rash, with 36 per cent of respondents indicating that they were unlikely (19 per cent) or definitely not (17 per cent) going surfing. It is worth noting that the issue that rated highest amongst respondents in the 'No way I'm going surfing' category is social or cultural and not biophysical. However, this fact has been intuitively known by surfers for quite some time, and underpins the work of the Groundswell Society. Society co-founder Glenn Hening remarks 'The Society intends to address the

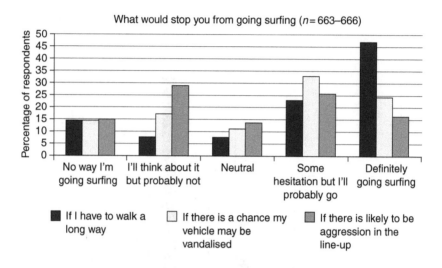

Figure 12.3 What would stop you from going surfing? 1.

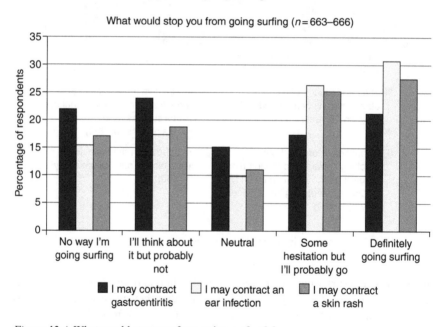

Figure 12.4 What would stop you from going surfing? 2.

fact that surfers get sick of each other more than they get sick from pollution' (Hening, in Barilotti, 2002, p. 10).

Least likely to keep a person out of the water is the fact that they would have to walk a long way to get to the surf, with 70 per cent of respondents indicating that they were likely (23 per cent) or definitely (47 per cent) going surfing. The second factor least likely to keep a surfer out of the water is the risk of contracting an ear infection with 57 per cent of respondents indicating that they were likely (26 per cent) or definitely (31 per cent) going surfing. Third least likely to keep a surfer out of the water is the chance of a surfer having his or her car vandalised with 57 per cent of respondents indicating that they were likely (33 per cent) or definitely (24 per cent) going surfing. Fifty three per cent of respondents indicated that they were likely (25 per cent) or definitely (28 per cent) going surfing even if there was a risk of them contracting a skin rash.

While biophysical factors are an obvious risk to surfers and are likely to cause many surfers to think twice about going surfing, especially in areas where the risk of illness is widely known, it is also interesting to note that aggression or 'surf rage' is a significant turn-off for many surfers. Surf rage is a term used to describe the actual and symbolic violence enacted by surfers attempting to protect 'their' break from unwanted outsiders (Nazer, 2004; Olive, 2015; Preston-Whyte, 1999, 2002; Young, 2000). When responses between Australian and US respondents were compared, what was remarkable was the similarity. However, US respondents overall indicated that they were less likely to surf if there they were at risk from contracting an illness. This may be attributable to US surfers' exposure to polluted water compared to generally better water quality for ocean bathing in Australia, however, detailed analysis has not been undertaken. With respect to the social indicators, the answers across the two sample groups are again quite similar with the one exception. Sixty per cent of American respondents indicated that they would definitely go surfing even if they had to walk a long way (22 per cent indicated that they would think about it but probably go) compared to only 39 per cent of Australian respondents (25 per cent indicated that they would think about it but probably go). Responses for NZ surfers ($n=40$), although quite low in number, possibly shed some light on the discrepancies. Fifty three per cent of NZ respondents indicated that they would definitely go surfing even if they had to walk a long way (30 per cent indicated that they would think about it but probably go). Perhaps Australian surfers have a different perception of what it means to walk a long way to get to a surf break compared to their US and NZ counterparts, or maybe we're just lazier?

Discussion: culture, ecology, and diversity

For many recreational surfers, the essence of their surfing experience remains a deeply personal and intrinsic experience, but one where enjoyment, self-improvement, fun and camaraderie have a strong role to play. This reminds us that individual experiences of surfing never stand alone, but are always cultural, contextual and spatial. This is despite the significant growth in participation and also the size of the surfing industry over the past 30 years and the perception of overcrowding at many surf breaks. The high level of importance given by survey respondents to the social indicators such as play, friendship and relaxation as well as a strong indication that survey respondents used their time surfing to be outdoors and connect with nature, suggests that surfing plays a varying but nevertheless important role in many participants' lives. As such, recreational surfers today can still be understood as 'serious leisure' participants (Lazarow, 2010).

Despite a defining image of surfers as 'lawless' and 'free', many surfers feel a strong current of 'personal responsibility' around social and environmental issues. This attitude highlights the potential of individualised nature or lifestyle sports like surfing, which are focused on individualisation and practices of consumption as key to cultural participation and identity (Wheaton, 2004, 2013). Then again, any discussion of environmental awareness is intimately linked with individualised motivations and relationships. That is, while surfers often emphasise the connections between surfing and environmental knowledge, this is largely limited to an individual level by surfers' overriding desire to access waves – lots of waves. Historically, levels of regular support for surfing-conservation organisations support this point (Anderson, 2014; Wheaton, 2007). However, this is distinguished from surfer-advocacy in the political rather than policy sphere/cycle, which has witnessed a growing level of participation over the past decade or so. Following this, there is a need to address the lack of empirical evidence to back up the conceptual claims relating to sport in a growing body of work about ecological sensibilities (Olive, 2015).

Of interest to the above point is the data around environmental factors that would stop individuals from going surfing. In terms of pedagogies of places and experiences, negative factors such as pollution and so on, can incite debates about environmental policies and engagement (Wheaton, 2007). An interesting recent example is in Australia, where there has been a spate of highly publicised (and sometimes fatal) shark attacks on surfers. These attacks have started debates around the culling of sharks for the safety of surfers and other ocean users, bringing into question the sense of connectedness to an ecosystem that relationships to the sea and coast are supposed to establish. We do not suggest that there is overwhelming support within the surfing community nor even majority support for the culling of sharks; simply that the renewed sense of consequence of shark presence has increased the intensity and controversy of the debate. The threat of shark attacks brings into very sharp and consequential relief the tensions between desiring a 'pure' or authentic experience of 'nature', and a desire to feel safe and secure in our experiences of the natural world. For surfers, these tensions of experiencing wildlife in its natural environment is heightened and much less controlled than some land-based sporting spaces.

The social and cultural surfing world is intimately connected with the natural world and must be considered as a 'surfing system' (like an ecosystem). Such a system includes industry and traditional, online and social media, which impact surfers' collective understandings of what surfing is and the role that surfing can play in life beyond the surf break and the beach. These collective understandings shape who is included and excluded, who are seen as 'natural' members of beach cultures and spaces, and whose knowledge of the coast is valued in community and policy discussions relating to environmental sustainability (Evers, 2008b;

Olive, 2015; Satchell, 2007, 2008). With the coast such a powerful part of Australian national identity, representations of surfers as coastal participants and activists must shape how the broader public makes sense of what the beach means in everyday Australian life.

Despite growing levels of female participation in surfing, the traditional embodiment of 'core' and 'local' are badges traditionally bestowed upon and worn by men – giving them the right to speak for place (Evers, 2008a; Olive, 2015). This raises some interesting issues. Most surveys of surfing effort suggest that a global participation rate of almost 90 per cent male, less so in a number of surfing's cultural bastions such as the Gold Coast, Noosa and Byron Bay (presumably in other places also), which attract greater female participation. While it is difficult to know what impact increased numbers of female participants are having on the cultures of lineups, Neil's results and Rebecca's research point towards constantly changing attitudes in the surf, which shift with various demographics of participants, including age, sex/gender, ethnicity, board type and ability level. While the role of sex/gender has been increasingly explored, the impacts of the global Western bias in surfing research based in Australia and the USA is an increasing issue of concern – that is, where and how indigenous and multi-cultural perceptions of place are excluded to privilege white, neoliberal understandings (Evers, 2008b; McGloin, 2005; Preston-Whyte, 1999, 2002; Roy and Caudwell, 2014; Satchell, 2007, 2008; Walker, 2005, 2011; Wheaton, 2013). Also, with the recent revelation that major surf brands are manufacturing in 'sweat shops', further attention needs to be paid to the tensions between surf industry production, environmental and social ethics, media representation and marketing, and the purported cultural values of surfers.

Conclusion

Surfing has definitely matured but in many respects has not come of age. The data collected by Neil suggests that surfers continue to see themselves and surfing culture as environmentally connected, with surfers developing a particular relationships to the environment and sustainability. However, not surprisingly, there are inconsistencies in these ideas, most notably embodied in the focus on individual access to waves and surf. While there is a cultural emphasis on individual experiences, ultimately, there is little evidence to suggest that the global surfing community has an overwhelming interest in or concern for the experiences and living conditions of others. Despite this, there remains an active cohort of surfers, community organisations and researchers, including ourselves, who continue to advocate for the potential of individual relationships and connections to place to lead to greater social and ecological sensibilities and activism. However, whether and how it is possible to imagine beyond surfers' immediate self interest in accessing waves is difficult to know. Like all systems of privilege, what surfers are willing to give up in order to secure better environmental and social conditions is difficult to know, and such decisions are often influenced by various social and cultural trends. Trade-offs such as higher priced gear in exchange for better working conditions for manufacturers, decreased wave power in exchange for harvesting wave energy, or decreased board performance in exchange for sustainable conditions are all possibilities that surfers may or may not be willing to accept. In these questions, the responses in this survey are both disappointing and heartening, highlighting the best and worst of Australian and USA surfing cultures. For both of us as surfers, this is not really a surprising revelation.

References

Anderson, J. (2009). Transient convergence and relational sensibility: Beyond the modern constitution of nature. *Emotion, Space and Society, 2*(2), 120–127.

Anderson, J. (2014). Surfing between the local and the global: Identifying spatial divisions in surfing practice. *Transactions of the Institute of British Geographers, 39*(2), 237–249.

Baker, T. (2008). An Open Letter to the Surf Magazine Editors of the World. *Kurungabaa, 1.*

Barilotti, S. (2002). Surf colonialism in the 21st century. *Surfers Journal, 11*(3), 50–52.

Bennett, R. (2006). *The Surfer's Mind: The Complete, Practical Guide to Surf Psychology.* Sydney: The Surfers Mind.

Booth, D. (1995). Ambiguities in pleasure and discipline: The development of competitive surfing. *Journal of Sport History, 22*(3), 189–206.

Booth, D. (1999). Surfing: The cultural and technological determinants of a dance. *Sport in Society, 2*(1), 36–55.

Booth, D. (2001). *Australian Beach Cultures: The History of Sun, Sand and Surf.* London: Frank Cass.

Booth, D. (2012). Seven (1+6) surfing stories: The practice of authoring. *Rethinking History, 16*(4), 565–585.

Boullon, L. A. (2001). *Surf Narratives: California Dreamin' on a new frontier.* (Master of Arts), California State University, Fullerton.

Challinor, S. (2003, 22–25 June). *Environmental Impact Assessment for Artificial Surfing Reefs: A United Kingdom Perspective.* Paper presented at the Proceedings of the 3rd International Surfing Reef Symposium, Raglan, New Zealand.

Evers, C. (2006). How to Surf. *Journal of Sport and Social Issues, 30*(3), 229–243.

Evers, C. (2008a). The Cronulla riots: Safety maps on an Australian beach. *South Atlantic Quarterly, 107*(2), 411–429.

Evers, C. (2008b). Rethinking gubbah localism. *Rethinking gubbah localism, 1*(1).

Ford, N. and Brown, D. (2006). *Surfing and Social Theory: Experience, Embodiment and Narrative of the Dream Glide.* Oxon, UK: Routledge.

Irwin, J. (1977). *Scenes.* Beverley Hills, CA: Sage Publications.

Johnson, J. C. and Orbach, M. K. (1986). The role of cultural context in the development of low-capital ocean leisure activities. *Leisure Sciences, 8*(3), 319–339.

Kampion, D. (1997). *Stoked! A History of Surf Culture.* Santa Monica, CA: General Publishing Group.

Kelly, J. (1973). *Surf Parameters: Final Report Part II Social and Historical Dimensions.* Honolulu: University of Hawaii James K. K. Look Laboratory of Oceanographic Engineering.

Kyle, G., Norman, W., Jodice, L., Graefe, A. and Marsinko, A. (2007). Segmenting anglers using their consumptive orientation profiles. *Human Dimensions of Wildlife, 12*, 115–132.

Lazarow, N. (2010). *Managing and Valuing Coastal Resources: An Examination of the Importance of Local Knowledge and Surf Breaks to Coastal Communities.* (PhD), Australian National University, Canberra.

lisahunter (2017). Becoming visible: Visual narratives of 'female' as a political position in surfing: The history, perpetuation, and disruption of patriocolonial pedagogies?, in H. Thorpe and R. Olive (eds), *Women in Action Sport Cultures: Identity, Politics and Experience*, pp. 319–348. London: Palgrave Macmillan.

McGloin, C. (2005). *Surfing Nation(s) – Surfing Country(s).* (PhD), University of Wollongong, Wollongong, Australia.

Manning, R. E. (1999). *Studies in Outdoor Recreation: Search and Research for Satisfaction. 2nd Edition.* Corvallis, OR: Oregon State University Press.

Miller, M. L., Auyong, J. and Hadley, N. (1999). *Balancing Tourism and Conservation.* Paper presented at the International Symposium on Coastal and Marine Tourism, Vancouver, British Columbia.

Nazer, D. (2004). The tragicomedy of the surfers' commons. *Deakin Law Review, 9*(2), 655–714.

Olive, R. (2013). Making friends with the neighbours: Blogging as a research method. *International Journal of Cultural Studies, 16*(1), 71–84.

Olive, R. (2015). Surfing, localism, place-based pedagogies, and ecological sensibilities in Australia. In B. Humberstone, H. Prince and K. Henderson (eds), *International Handbook of Outdoor Studies* (pp. 501–510). London: Routledge.

Olive, R. (2016). Going surfing/doing research: Learning how to negotiate cultural politics from women who surf. *Continuum: Journal of Media & Cultural Studies*, doi: 10.1080/10304312.2016.1143199.

Olive, R., McCuaig, L. and Phillips, M. G. (2015) Woman's recreational surfing: A patronising experience. *Sport, Education and Society, 20*(2), 258–276.

Ormrod, J. (2007). Surf rhetoric in American and British magazines between 1965 and 1976. *Sport in History, 27*(1), 88–109.

Pearson, K. (1979). *Surfing Subcultures of Australia and New Zealand.* Brisbane, Australia: University of Queensland Press.

Preston-Whyte, R. (1999). *Contested and Regulated Seaside Space at Durban.* Paper presented at the International Symposium on Coastal and Marine Tourism: Balancing Tourism and Conservation, Vancouver, British Columbia.

Preston-Whyte, R. (2002). Constructions of surfing space at Durban, South Africa. *Tourism Geographies, 4*(3), 307–328.

Roy, G. (2013). Women in wetsuits: Revolting bodies in lesbian surf culture. *Journal of Lesbian Studies, 17*(3–4), 329–343.

Roy, G. and Cauldwell, J. (2014). Women and surfing spaces in Newquay, UK. In J. Hargreaves and E. Anderson (eds), *Routledge Handbook of Sport, Gender and Sexuality* (pp. 235–244). Abingdon, UK: Routledge.

Satchell, K. (2007, 6–8 December). *Shacked: The Ecology of Surfing and the Surfing of Ecology.* Paper presented at the Sustaining Culture, Annual Conference of the Cultural Studies Association of Australia, Adelaide.

Satchell, K. (2008). Reveries of the Solitary Islands: From sensuous geography to ecological sensibility. In A. Haebich and B. Offord (eds), *Landscapes of Exile: Once Perilous, Now Safe* (pp. 97–114). Bern: Peter Lang.

Stebbins, R. A. (1979). *Amateurs: On the Margin Between Work and Leisure*. Beverley Hills, CA: Sage Publications.

Stranger, M. (2011). *Surfing Life: Surface, Substructure and the Commodification of the Sublime*. Farnham, England: Ashgate.

Walker, I. (2005). Terrorism or native protest? The Hui 'o he'e nalu and Hawaiian resistance to colonialism. *Pacific Historical Review, 74*(4), 575–601.

Walker, I. (2011). *Waves of Resistance: Surfing and History in Twentieth-century Hawaii*. Honolulu, HI: University of Hawai'i Press.

Warshaw, M. (2004). *The Encyclopedia of Surfing*. Melbourne, Australia: Penguin Books.

Wheaton, B. (ed.). (2004). *Understanding Lifestyle Sport: Consumption, Identity and Difference*. London: Routledge.

Wheaton, B. (2007). Politics and the Beach: Environmental Activism in Surfers Against Sewage. *Leisure Studies, 26*(3), 279–302.

Wheaton, B. (2013). *The Cultural Politics of Lifestyle Sports*. Oxon, England: Routledge.

Young, N. (2000). *Surf Rage: A Surfer's Guide to Turning Negatives into Positives*. Angourie, Australia: Nymboida Press.

Young, N., McGregor, C. and Holmes, R. (1994). *The History of Surfing. Revised Edition*. Sydney, Australia: Palm Beach Press.

13 Simulating Nirvana

Surf parks, surfing spaces, and sustainability

Jess Ponting

Introduction

On 18 December 2015 11-time world surfing champion Kelly Slater took to Instagram to release a three minute 40 second video of the Kelly Slater Wave Company (KSWC) test facility in the rural central valley of California. Within an hour the video had been seen 250,000 times. Within four days more than four million people had viewed it on Facebook alone (Pamer, 2015). The waves generated were significantly better than any previous human made wave and news outlets around the world favourably compared it to some of the world's best surf breaks. Surf tourism literature has addressed the social construction of idealized surfing spaces (referred to as Nirvanification) and the implications of this for the sustainability of surfing destinations (Ponting, 2008; Ponting and McDonald, 2013; Preston-Whyte, 2001, 2002), but not in the context of the surf park industry. In addition to KSWC two wave generation technologies are already in operation, another two are slated to open in 2016 and a fifth is reportedly close to closing several deals. The industry has the potential to create major lifestyle, recreation and tourism attractions and radically transform the world's access to surfing experiences. Despite the potential impact of surf parks on the $130 billion a year (O'Brien and Eddie, 2013) surf recreation, tourism, media and consumer goods industries and the communities, lives, and environments they touch, surf parks are an under studied phenomenon. This chapter aims to establish the landscape of the surf park industry and begin to explore the implications of constructing wholly artificial surfing spaces from a sustainability perspective.

A brief history of surf parks

The history of human made waves has a surprisingly long history and begins in the Venus Grotto of the 'mad king' Ludwig II of Bavaria in 1877. Designed by Fidelis Schabet, the grotto was installed at Linderhof castle and was equipped with electric lighting in multiple colours, central heating, and a wave machine that directly applied electricity to create what were essentially ripples across a small artificial lake (Blunt, 1970). In the second quarter of the twentieth century surf pools began to be built with more frequency to provide enhanced swimming experiences. This began in 1903 when German Hofrat Höglander patented a hydraulic flap system to create breaking waves in a pool. The Undosa-Wellenbad was presented as a medically efficacious surf bathing experience at the 1912 International Hygiene Exposition in Dresden. With 7,500 paying customers utilizing the pool each day, the surf bath also had great economic potential (Westwick and Neushul, 2013). Another surf pool was built in 1927 as part of an expansion of the Gellert Baths (which were originally built and opened in 1918) in Budapest, followed in 1929 by a pool in Berlin, the Empire Pool in Wembley, London in 1934, and the Pallisades Amusement Park in New Jersey in the United States (Warshaw, 2004; Westwick and Neushul, 2013).

Warshaw (2004) estimated that somewhere in the realm of 500 wavepools were built for enhanced swimming experiences in the twentieth century. The crossover with surfing began in 1966 in the Japanese Summerland Wavepool known as the Surf-a-Torium which set aside a quarter of each hour for surfers to ride weak, spilling, 2–3 foot waves. In 1969 the two million dollar Big Surf pool in Tempe, Arizona opened, a 300×400 foot pool utilizing a 40 foot water drop system to create a chest-high spilling wave across the pool. In the early 1970s the Offshore Technology Corporation (OTC), founded by ex-Navy submarine commander Frank Biewer in 1968, was building wave-testing facilities for oil platforms and saw an opportunity to develop surf pools. Eventually a new company, Surf Tech, spun off from OTC and began building wavepools in 1984 beginning with the rather limp Aquaboggan water park in Saco, Maine followed by Dorney Park Wildwater Kingdom, Allentown, Pennsylvania in 1985 (Lowry, 1985). With the latter an attempt was made to mainstream surf pool surfing with the World Professional Inland Surfing Championships, but the event did not inspire competitors or the viewing public (Warshaw, 2004).

Surf pool quality improved significantly with the second generation of 'water drop' surf pools heralded by pool engineering firm Barr and Wray's Typhoon Lagoon at Disney World Orlando in 1989 (John Luff, pers. comm., 2016) – a style of wave generation that continued to set the standard for affordable surf pool installations as part of larger water park attractions until the arrival of third generation wave pools in the second decade of the twenty-first century. In 1990 San Diego's Tom Lochtefeld began experimenting with 'sheet wave' technology, coming to market with the first Flowrider in Texas in 1991 and following it up with the first Flow Barrel in Norway in 1993 (Waveloch, 2016). While by Lochtefeld's own admission sheet wave attractions do not provide an authentic surfing experience, they are fun. By 2015 Waveloch had installed more than 200 sheet wave attractions in 35 countries. Also in 1993 in Miyazaki, Japan, the 300 million US dollar Ocean Dome was opened as part of the Sheraton Seagaia Resort. A massively expensive wave generating system developed by Mitsubishi Heavy Industries (which also builds nuclear power plants) powered what were to remain the best man made waves for more than 20 years. The sheer cost of the system and a two minute interval between waves made a return on investment an almost impossible dream. Despite the unprecedented wave quality, Ocean Dome eventually closed in 2007.

The surf park industry was dealt another blow in 2008 with the failure of the Ron Jon Surf Park. The project brought together many of the best minds in surf park development including the company Artificial Surfing Reefs that had mapped the bathymetry of many of the world's best surf breaks and hoped to be able to re-create them with a programmable robotic pool floor. A highly successful pre-development phase proved there was demand for the Ron Jon Surf Park, established financial viability for the proposed model, and raised sufficient capital ($10M) to execute the project. Unfortunately the pool itself was a spectacular failure and never opened to the public. Surfing Magazine's Jimmy Wilson summed up the attitude of many in response to the failure of the Ron Jon Surf Park.

> I guarantee you investors have taken note of this hideous failure. They took top-notch technology and millions of dollars to create a contraption that is easily shown up by neighbor Typhoon Lagoon's wave pool, which wasn't even built specifically for surfing and was constructed in 1989!
>
> (Wilson, 2008)

The Ron Jon debacle combined with the global recession led to investment in major surf park developments all but drying up in the immediate term with the exception of Murphy's Waves' 2012 Wadi Adventure surf pool in Al Ain in the United Arab Emirates. Wave technology companies did not, however, stop the process of research and development. Indeed, Wadi Adventure is likely to be the last second generation water drop pool designed as a loss leading anchor attraction. Third generation pools capable of producing very high-quality waves, intermediate, and learner waves, in some cases simultaneously are now a reality. Five main companies are competing in the third generation pool space utilizing two kinds of technology to build surf parks as profitable stand-alone attractions. Wavegarden, Kelly Slater Wave Co, and Webber Wave Pools' technologies are all based on slightly different applications of the same basic concept: mechanical foils being dragged through the water. Alternatively, Surfloch and American Wave Machines both use sequentially fired pneumatic caissons to generate waves. Some of Surfloch's designs also incorporate bathymetric features on the pool floor. The first of these to market was Surf Snowdonia, a Wavegarden project in Conwy, North Wales that opened to the public on 1 August 2015. The facility attracted 15,000 visitors in its first month of operation and staged a successful major international competition in its second month. A breakdown in September 2015 caused a six-day closure before a major mechanical failure and tear in the lake liner forced the park to close for the winter eight weeks early. Though it does produce a shoulder-high spilling wave every 90 seconds, Wavegarden failed to deliver to Surf Snowdonia the six foot plunging wave every 60 seconds it had promised in pre-opening media releases (Temperley, 2015). Surf Snowdonia reopened on 19 March 2016 with a marginally improved product as a result of work done on its northern distributive shores to hasten the dampening of wave energy. A second, larger, Wavegarden project is due for completion in

2016 in Austin, Texas. American Wave Machines has projects under construction in Sochi, Russia and in Meadowlands, New Jersey. The former is due to open in 2016 and the latter in late 2017. Surfloch has two relatively small-scale projects under construction in Rotterdam, Netherlands and in Lisbon, Portugal, and a third full-scale installation outside Bristol in the United Kingdom that has passed through the government approvals process. Kelly Slater Wave Co, as mentioned previously, has produced a full-scale prototype, it does not yet appear to be commercially viable based on throughput and the lack of simultaneous intermediate and beginner options. Webber Wave Pools is yet to develop a working prototype or break ground on a fully funded project but has a conceptual product that is very appealing.

Social and political landscape of factors influencing surf park development

There are an estimated 3.3 million surfers in the US and 35 million surfers worldwide (O'Brien and Eddie, 2013) growing at 35 per cent annually (WSL, 2016). In the United States the average surfer is a highly educated 34 year-old male earning $75,000 a year who owns four surfboards, travels 20 miles to surf for 2.5 hours on 108 separate occasions annually and pays between $59 and $100 dollars for each surf on items including food, gas, rental equipment, lodging, and/or merchandise. US surfers inject US$2–5 billion directly into coastal economies around the United States (Wagner *et al.*, 2011). Beyond the US, the International Surfing Association has 97 member countries. Surfers in each of these countries generate a significant and growing economic impact locally and internationally. Surf tourism takes place in at least 162 countries (Martin and Assenov, 2012) and incorporates all levels of luxury and service. The World Surf League estimates the global audience for competitive surfing to be 120 million (WSL, 2016). Anecdotally, most international surf tourists are based in the US, Australia and New Zealand, Brazil, Japan, and Europe (UK, France, Spain, Portugal) though emerging markets include Russia and Central and South America. A recent study showed that 91 per cent of surfers from a range of countries had taken an international surf trip in the past five years. Of those, 82.1 per cent had taken more than two surf trips, almost 40 per cent had taken more than ten surf trips, and almost 20 per cent had taken more than 21 surf trips in the past five years (Barbieri and Sotomayor, 2013) – this during a period of global economic recession. The global surf industry, inclusive of surf tourism, has been estimated to be worth $130 billion annually (O'Brien and Eddie, 2013). Surfers are highly motivated, mobile, and reactive travellers – they follow the surf and are willing to spend a great deal of money for the chance to ride uncrowded quality waves. Data presented at the 2015 Sustainable Stoke Conference showed that 11 per cent of 3,000 surfers sampled would travel internationally to ride a high-quality surf pool (Ponting, 2015).

Natural surf breaks generate significant economic activity. Australia's Gold Coast alone generates $819.9 million each year from surf tourism (Gold Coast

City Council, 2009). A run of swell in February 2016 was estimated to have delivered a $20 million boost to the economy of the City of Gold Coast through additional surf tourism (McElroy, 2016). Uluwatu in Bali, Mundaka in Spain, Mavericks in northern California, and Stradbroke Island in Australia generate $8.4 million (Margules, 2011), $4.5 million (Murphy and Bernal, 2008), $23.8 million (Coffman and Burnett, 2009), and $20 million (AEC Group, 2009), per annum respectively. In addition to calculations on the local and regional economic impact of individual surf breaks, the hedonic price method has been used to establish that after controlling for proximity to the beach, ocean views, the specific characteristics of the homes, and neighbourhood effects, a home adjacent to a quality surf break is valued at approximately $106,000 higher than an equivalent home a mile away (Scorse *et al.*, 2015). This may have implications for real estate adjacent to surf pools, and conceivably value could be added to residences, timeshares, and resort or hotel rooms through the addition of a quality surf pool.

Adding to market readiness for a surf park industry is the fact that the global surfing population is large and growing quickly at a time when threats to the surfing environment and experience have never been more pronounced and immediate. These include crowding, pollution, the impacts of climate change, and perceptions (and real evidence) of growing numbers of shark attacks. The latter was highlighted by the 2015 attack on three-time world surfing champion Mick Fanning during the final of a World Surf League contest in South Africa, broadcast live to millions of fans and subsequently viewed online more than 20 million times. The east coast of Australia has suffered 12 attacks in the past 12 months alone, three fatal, many in areas unused to shark sightings and activity. Unreported minor attacks and bumpings are anecdotally taking place everyday, and unofficial reports of sharks chasing surfers out of the water without an actual attack are very common. The governments of Reunion Island and the New South Wales state government in Australia have been forced to implement measures to protect surfers from shark attacks. In both places the mindset of surfers has been significantly influenced by the threat of shark attacks. Surf-related tourism has suffered and local surf retailers and surf schools have also witnessed declines. In the case of Ballina, Australia, the town most impacted by the recent increase in shark attacks, local government officials have called for the construction of a surf park to ensure the town does not lose its appeal as a surfing destination and to protect local surfers (Lollback, 2015).

Many prominent surfers have advocated for human interventions to address the growing imbalance between an oversupply of surfers and an undersupply of waves. The 1978 world surfing champion and former CEO of the Association of Surfing Professionals, Wayne Rabbit Bartholomew, called for a balance between natural and man-made waves to address growing demand and static supply on Australia's Gold Coast (Bartholomew, 2015) and such considerations have been included in the 2016 Gold Coast Surf Management Plan (City of Gold Coast, 2016). The 1977 world surfing champion Shaun Tomson is an outspoken advocate for human interventions on natural coastlines to improve surfing amenity and create local economic opportunities (Tomson, 2013). Kelly Slater, perhaps the most famous surfer of all time, has invested his own money and name in the development of Kelly Slater Wave Co.

Under the auspices of the International Surfing Association, and in particular Fernando Aguerre, its charismatic president and co-founder of successful surf brand 'Reef', there has been an aggressive push for surfing to be included in the line-up of sports for the summer Olympics. Building the case for surfing has involved an aggressive expansion of ISA member countries which at the time of writing sit at 97, Norway being the most recent addition. A significant component of the calculus for Olympic inclusion has been evening out the playing field in terms of the element of chance that the ocean introduces to competition, and allowing countries with no surfable coastline to host surfing. For these purposes the development of a high-quality surf pool has been considered a prerequisite to a successful Olympic bid. In September 2015 the Japanese Olympic Committee recommended to the IOC that surfing be included in the 2020 Tokyo Olympics. The final decision will be made by a full sitting of the IOC during the 2016 Olympics in Rio de Janeiro, Brazil. Interestingly, despite the success of Surf Snowdonia (and perhaps because of the mechanical issues it suffered) the Japanese committee has recommended that surfing take place at an ocean venue rather than a surf pool (Inertia, 2015a) – though this may be subject to change, particularly in light of the successful trial of KSWC's test facility.

An additional barometer of the state of the surf park industry was the founding of the Surf Park Summit (SPS). First held in 2013 and co-founded by the Center for Surf Research at San Diego State University and Surf Park Central (an industry information consolidation organization), the SPS aimed to be an industry accelerator and attracted more than 300 attendees including representatives of four of the six viable technology providers and senior management of a range of surf corporations. A second Surf Park Summit is planned for September 2016 with an expected 500 attendees and an international cast of presenters and panellists including technology companies, financiers, specialist contractors, athletes, developers, operators, media, academics, and surf related NGOs.

The holy grail effect

There is a clear relationship between the number of surfers at a surf break and the number of waves available to any one surfer (Ponting, 2008). Studies show that as crowding of outdoor recreation resources increases, visitors with lower crowding thresholds and higher daily spending patterns are displaced by those with higher crowding thresholds and a concomitant lower daily spend (Budruk et al., 2008; Manning, 2007; Navarro Jurado et al., 2013; Vaske and Shelby, 2008). The commercial surf travel industry has intuitively understood these issues and beginning with the world's first surf resort at Garajagan on the eastern tip of Java in 1979 establishing exclusive use of surf breaks has been considered the industry's Holy Grail (Ponting, 2008). The world's second surf resort, Tavarua Island Resort in the Mamanuca group of islands in Fiji similarly sought to establish exclusive access to two of the world's best waves (Ponting and O'Brien, 2014a). A number of resorts in the Maldives exploited the 'boundary regulation' law 2012/R-7 (Ministry of Tourism Arts and Culture, 2012) that allowed exclusive resort use extending 800 m out from island vegetation lines to claim exclusive access and management rights over surf breaks within these zones (Ponting, 2014). Three separate attempts at regulation in Indonesia's Mentawai Islands have attempted to provide some level of security to investors in resort developments with regard to crowding and access management (Ponting and O'Brien, 2014b). A loose association of resort owners in Indonesia's Telo Islands attempted to establish a management plan to limit access in concert with local government agencies (Ponting and Lovett, 2015). These have been incredibly contentious initiatives and almost all attempts have failed because of stakeholder resistance, a lack of appropriate legislation to regulate surf breaks, or a lack of political will. With management and regulation rights residing unambiguously with surf park developers, surf parks are in the enviable position of being able to provide a guaranteed wave count for surfers without the accompanying political controversy. For example, Surf Snowdonia is able to guarantee its advanced surfers 12–13 hassle free waves in a one hour session.

In addition to wave count, surf parks are able to guarantee wave size, shape and quality. Surf breaks are of poor quality far more often than they are breaking perfectly. High-quality waves require the alignment of a complex array of factors including swell size, period, direction, tide, local wind direction and strength, and bathymetry. Many of these factors are seasonal and international surf tourism demand shifts accordingly. In recent years seasonality has been amplified by the development of sophisticated surf forecasting technologies and services including live streaming webcams at thousands of surf breaks around the world. The result has been the rise of 'hyper-seasonality' based on the ability to accurately forecast surf conditions as much as two weeks in advance. This has significantly impacted many surf tourism destinations as booking lead times dwindle to almost nothing and occupancy rates in the surf season lurch from over capacity to empty and back again at the whim of long-range surf forecasts. Surf pools remove most of these variables.

Nirvanification

Ponting (2008) and Ponting and McDonald (2013) outlined a theory of Nirvanification that details the apotheosis of surfing places into idealized, largely mythical surfing spaces in the context of international surf tourism. This work, and the work that informs it, has implications for understanding what the key symbolic elements of an entirely artificial surfing space are likely to be and how they may interface with local cultures, environments, and broader sustainability concerns. For example, Edensor (1998: 42) proposed a continuum from 'enclavic spaces' such as integrated resorts, theme parks, shopping malls, and for our purposes surf parks at one pole, and at the other 'heterogeneous spaces' (for example a beach with competing user groups: surfers, fishers, joggers, swimmers, sun bathers, boaters, etc.) at the other, based on Foucault's hetero-topias: 'the juxtaposing in a single real place (of) several spaces, several sites that are in themselves incompatible' (Foucault, 1986: 25). While some have assumed that space is reliant on a specific physical place (Cunningham, 2006; Johnston, 1992; Tuan, 1975), Ponting and McDonald (2013) argued that com-mercial surfing tourist spaces may manifest as a universal cultural space wher-ever the requisite symbolic elements are present. In this way surfing destinations can become 'disembedded' from their geographic and cultural context (Giddens, 1990). Geographers have referred to these kinds of disem-bedded tourist spaces as 'free floating signifier[s]' (Hopkins, 1998: 65), or as 'travelling representations' (Davis, 2005: 607). In the context of commercial surf tourism Ponting (2008) and Ponting and McDonald (2013) present Nir-vanification as a four-phase dialectic process through which the disembedding of surfing space occurs.

The first phase involves the social construction of Nirvana. In the case of Ponting and McDonald's (2013) empirical work this was based on the presence of four symbolic elements: perfect waves, uncrowded conditions, cushioned adventure (a sense of adventure despite considerable levels of comfort and safety provided by the tourism industry), and a pristine tropical environment. In the second phase the symbolic elements of Nirvana are threatened. As tourists engage in the co-performance of surf tourism, the pre-packaged meanings of place provided to them by the surf media and surf tourism industries are renego-tiated based on personal embodied experiences of place. For example, tourists may find that the waves are not perfect all the time, the breaks may be crowded, local people may be unwelcoming, and the environment may be suffering as a result of their presence in the destination. In the third phase of Nirvanification the surf tourism industry responds to threats through the deployment of myth – a typical response of the powerful to competing constructions of reality (Berger and Luckmann, 1966). For example, to help tourists rationalize experiences with local people struggling to feed their families adjacent to their luxury surf resorts or charter boats, the surf tourism industry in Indonesia's Mentawai Islands prop-agated myths that local people were content with their lives and had no interest in participating in surf tourism. Further, the myth posited that to involve local communities in surf tourism would inevitably lead to their social, cultural, and moral decline (Ponting, 2009; Ponting and McDonald, 2013). These 'Nirvanic myths' were largely successful in 'nihilating' (Berger and Luckmann, 1966: 12) the threats to Nirvana. However, in the fourth phase of Nirvanification altern-ative discourses emerge from local voices and importantly as a result of tourists' embodied experiences of place that contest and contradict Nirvanic myth. The emergence of a robust alternative discourse that can withstand nihilation attempts through Nirvanic myth deployment has the power to destabilize Nirvana.

If Nirvana is co-performed by tourists and locals in a dialectic process of production and consumption, embodied experiences of place and people that transcend symbolic representations have the power to transform the performance of Nirvana to reflect new understandings and meanings.

(Ponting and McDonald, 2013: 428–429)

In contrast to the heterogeneous space of a re-embedded surfing destination, surf pools are the epitome of enclavic space. Entirely artificial, and constructed for commercial purposes, the tightly controlled enclavic spaces of surf parks have the potential to replicate issues found in other such spaces. Under the guise of freedom and escape, these kinds of spaces are subject to intense social control, leaving little room for local understandings of space. Large capital investments and the international tourist standards, aesthetics and mythical disembedded narratives of space that often accompany them can subjugate local styles, culture, and environmental concerns. A number of questions arise in the context of the emerging surf park industry. What are the most important factors in an artificial surfing space? Which elements will be subject to reality maintenance? What form will threats, myths, and resistance take in the context of artificial surfing spaces? These questions will be considered in the following sections.

Nirvanification and designer surfing spaces

In-depth semi-structured interviews were conducted with two surf park technology developers, Greg Webber and Tom Lochtefeld, two surf park attraction developers, Nick Hounsfield and Andy Ainscough, and John Luff, founder of Surf Park Central, and co-founder of Surf Park Summit and consulting firm Surf Park Solutions. Interviews varying between 40 and 90 minutes were conducted via Skype or in person and in both instances were video recorded, then transcribed. In addition, an online survey was conducted in May/June 2015 that used snowball sampling to reach surfers through social media. Participants were incentivized to complete the survey by being entered into a draw to win a free surfboard of their choice from popular surfboard manufacturer Firewire. Three thousand and fifty-six surfers from 67 countries completed the survey which covered a broad range of academic territory from issues of sustainability, use of technology, and attitudes towards and potential usage patterns of surf parks. Only relevant surf park data will be considered here. It should be noted that the survey was completed before Surf Snowdonia opened and before the Kelly Slater Wave Co video was released. These two events showcased very high-quality surf pool waves and featured widely viewed endorsements of the wave quality by high-profile professional surfers. It is possible that the needle may have shifted a little for surfers who at the time of the survey were questioning surf park wave quality.

Interestingly surf park technology and attraction developers have divergent but interrelated ideas on what the key symbolic elements of surf park space might be. Australian Greg Webber, of Webber Wave Pools, while acknowledging that elements like aesthetics and human traffic flow will be important opined that the fulcrum for success in the industry can be distilled down to a single surf pool attribute – wave quality.

This question is absolutely fundamental to the success of this whole industry. The critical thing that surfers look for is really just the wave itself. It's so narrow it's ridiculous. For decades now people have had moments in which they've done a turn or got a barrel and then gone back to school or gone to work and told their mates. Not one of those moments would be damaged by the guy saying 'but I'm so disappointed that I was doing it in an urban environment'. Now that doesn't deny the fact that of course when guys come back from an experience overseas, that the entire experience is infused into them, but I don't think that it's actually transformative to the same degree that the actual wave itself is … the experience of doing it is fundamentally a transformation of your sense of place, because while you're doing it time is distorted. Place is irrelevant in the moment. It will come down to are the waves good or are they not good?

Webber makes a powerful case that riffs on Csikszentmihalyi's (1975) notion of 'flow' in which a person's focus of attention is entirely on the task at hand at the exclusion of other sensory inputs resulting in a state in which the passing of time is difficult to gauge. In addition Webber noted elements of the holy grail of surf break management previously discussed.

We won't have to have all that [crowding and hassling] crap. All you'll be doing is queuing up for a few minutes and then you know for sure that no one is hassling you at all. The wave is yours to enjoy, to be at peace with it or rip the thing to pieces. We don't fully yet appreciate how nice that will be. The absence of hassle will be probably the most important experiential component.

Tom Lochtefeld founder and CEO of Waveloch and Surfloch noted three vital components of a surf park. The first, and similar to Greg Webber's position, is the presence of a very high-quality wave. However, Lochtefeld's reasoning differs from Webber's in that the primary barrelling wave is there to attract highly-skilled surfers who will lend credibility and cultural capital to the surf park and act as inspiration to the aspirational surfers who will serve as the facility's cash cows by paying similar rates to the skilled surfers for quality instruction en masse. The second component is purposefully scaffolding surfing experiences in such a way as to progress first-time surfers towards the end goal of surfing the primary wave. Finally, Lochtefeld explained that a full lifestyle experience based on surfing culture and focused around acquiring the skills required to surf the primary wave is necessary to keep young aspirational surfers and their parents coming back.

Ninety per cent of the people that will participate won't be able to surf that perfect wave that will be on the outside barrelling. You need to appeal to the 90 per cent and make an aspirational lifestyle development that is much much more than the wave itself. But with that said you need the wave to begin with. Without having the perfect wave you won't get the buy-in from the core surfer who you want. That pro with a very high level of expertise. You have to have these scales of waves that are going to allow the aspirational aspect to create the money to pay for everything. That's the bottom line. But you've also got to get the photo.... If you don't have the 'wow shot' you'll never get that buy-in. There are different levels of how you create that. Some of it's fabricated, some of it's real. The more real you can make it the more legitimate and the longer trajectory you would have in your development. For me I want to have that but at the same time be safe. Be economically viable, and have something that can appeal to those masses. In the design of that facility, you have to throw everything into the mix to be a success.

Surf park developer Nick Hounsfield, who considers his pending project, 'The Wave, Bristol' to be a public service facility with positive heath and wellness implications for many, noted a number of challenges to creating an authentic feeling surfing space in an inland destination. Hounsfield worries that the surf park experience may become disembedded from important cultural experiences and reference points of the sport and specific locations. He believes that many of these issues can and should be considered the design, operation, and construction of surfing space at surf parks.

The trick will be designing it so its looks like it hasn't just landed there and is sticking out of the ground.... The design challenge is creating that reveal moment where you would go over the top of a sand dune and suddenly see waves. That sense of search and getting lost in the landscape is a really important part of what you lose by having an inland surfing lake, but potentially if you do it right you could bring that back in a slightly quirky way.

As the only developer with an active third generation commercial surf pool, Managing Director of Surf Snowdonia, Andy Ainscough, is uniquely positioned to comment on how the key concerns for creating an authentic surfing space are playing out. Ainscough noted the importance of having a 'surfer at heart' involved in creating the experience to, in essence, 'gut check' the commercial setting and product offering. In terms of maintaining the authenticity of the experience beyond the embodied experience of actually surfing the wave, Ainscough has found that a staff of core surfers lends itself to recreating the kind of interactions surfers in surfing communities or in surf tourism destinations might expect: surfing during the day and dissecting the day's surfing and telling surf stories by night.

Keep someone who is a surfer at heart right through the development. You've got to have that business sense to intuitively know what feels a little bit stupid or what's unreasonable to expect a surfer to do. Right throughout the build looking out over the Snowdonia mountain region we could see that if you were paddling out here it would be a great experience. It doesn't mean you can't have one in a city centre. I think that would still work but I don't think it would have quite the same soothing effect as it does with forest surrounding and a mountain backdrop like we have. Once you have a surf park and you employ 40 surf instructors and a similar number of life-guards it keeps that vibe automatically where people surf during the day and go in the bar at night. You automatically capture the essence of surfing really just by the people you end up taking on to work here.

John Luff's notion of surf park space is inextricably meshed with the need to accommodate the broader markets that surf parks must serve in order to succeed. He noted four symbolic elements: accessibility in terms of price *and* distance from major population centres, wave size and quality range to cater for multiple audiences, wave frequency to ensure viability and accessibility, and safety to appeal to those with fear of the ocean and what resides in it. Luff's position on wave size and quality echo that of Tom Lochtefeld on the imperative of through-put and is fairly self explanatory. On accessibility and safety he noted:

> Accessibility should be viewed in both the geographical and economical sense of the word. While there are certainly many people around the world willing to travel great distances and spend considerable sums of money on high-quality surfing experiences, relying solely on this market significantly limits the customer base.... Safety should not be overlooked as this is vital to attract the first-time surfers, beginners, and those that have generally been averse to ocean-based surfing experiences throughout life. An incredible amount of people around the world have stated a variety of reasons for not trying surfing even though they believe it looks fun and they'd like to try it. These reasons include, but are not limited to, fear of sharks, sting-rays, jelly-fish, rip currents, murky water, poor water quality, and uncontrollable forces of nature in the ocean. Parents of young children and teens are comforted and assured knowing that their kids are in a controlled ocean-like environment under the close supervision of lifeguards and coaches.

In-depth interview participants raised a number of important points in outlining their ideas about what the key symbolic elements of an artificial surfing space will be. There are close parallels with the symbolic elements described by Ponting (2008) and Ponting and McDonald (2013). However, additional nuance is added to the idea and utility of perfect waves in a purely commercial setting, and there is a shift from a focus on crowding to a focus on wave count as a metric of experience quality. The elements of exploration, freedom, and adventure remain but are noted to be elements that must be manufactured and specifically designed into facilities and operations and also need to be balanced with the need to cater to the fears of those frightened of the ocean. Similarly, creating an authentic social and cultural experience is something that requires active design and management consideration. The setting remains important but is not limited to pristine natural environments. Interestingly, accessibility has become a key symbolic element in contrast to Ponting's (2008) Nirvana, which in many ways is reliant on inaccessibility for its longevity. The following six points summarize they key symbolic elements of surf park space.

1   perfect waves
    a   distort time and remove the importance of place
    b   vital for creation of an aspirational lifestyle development
    c   variable for different user groups
2   guaranteed wave count (wave frequency)
3   cushioned adventure: building a sense of exploration, freedom and adventure into the experience while providing reassuring levels of safety service
4   authenticity from a social/cultural perspective
5   the setting: preferably impressive in terms of either natural surroundings or built design
6   accessibility (cost and distance).

Survey data suggests that surf park visitors will be more tightly focused on the controllable elements of a surfing experience promised by a surf park. Wave quality, a guaranteed wave count, length of ride, and the ability to choose wave height were by far the most important elements in a surf park experience amongst survey participants. It should be noted, however, that all the respondents are surfers. As Tom Lochtefeld explained and Greg Webber, Andy Ainscough, and John Luff corroborated, while vital to developing an aspirational market, established and competent surfers will not be the main economic drivers of surf parks: beginners and learners will be. While no data exists on the preferences and priorities of beginners, one might assume that factors including waves suitable for learning, safety, quality of instruction, training technologies, physical comfort etc. may be more important than they are for experienced surfers.

Threats

Survey participants who indicated they would use a surf park (78 per cent, $n=2134$) were asked to provide an open-ended explanation of their biggest concern when visiting a surf park for the first time. A number of themes emerged including: safety and injury, falling early on a wave and wasting it, an aversion to spectators and 'scenesters', what surfboard to ride, water quality, mechanical breakdowns, environmental impacts and sustainability (specifically, concerns about land use, power generation and water usage), and the authenticity of the surfing experience. The overwhelming majority of responses, however, were concerned with elements of crowding (e.g. wait times, getting enough waves, having to hassle for waves), cost, and wave quality (power, size, length, shape, frequency). In addition to those identifying as potential surf park users, the 22 per cent of participants ($n=598$) who indicated that they would not use a surf park were asked why. Their responses mirrored the concerns of the group that would use surf parks. A small number noted that they (a) didn't want to surf with an audience, and (b) didn't want to use a facility they considered to be environmentally unsustainable. Almost all other responses were related to quasi-spiritual (and explicitly spiritual) benefits derived from ocean surfing. The following responses to the survey are representative.

> Surf pools can never be authentic. Surfing is about connecting with nature and its energy.

> I go surfing not only to surf, but to be in a natural environment of which I am a small element. It reminds me to be humble, to appreciate the beauty of nature, and it calms my soul. A concrete, highly chlorinated, machine-enabled wave pool would do nothing more than reduce a precious resource and rejuvenating experience to a simple commodity.

> I enjoy surfing because I enjoy the ocean, not because I want to catch waves.

Tackling this spiritual divide between an ocean surfing experience and a surfing experience outside of the ocean is likely to be one of the toughest challenges facing the surf park industry. The solution may be more of a product positioning exercise, or a marketing exercise than a design aesthetic or even an operational challenge. None of the developers currently in the market are suggesting that their products can or even want to compete directly with surfing in the ocean. All acknowledge that the ocean holds an unbeatable spiritual appeal to surfers, themselves included. Andy Ainscough, for example, described his relationship with Surf Snowdonia as 'training for when there is good surf'. John Luff reiterated this idea: 'Surf parks are able to offer the perfect training grounds for the ocean. Learn in the pool, graduate in the ocean.' It may be important to highlight fitness and training messages rather than positioning surf pools as an ocean replacement on a spiritual level for the 22 per cent of the potential market with concerns about the spiritual divide.

   The enclavic nature of surf park space enables the Holy Grail effect from a management perspective but also comes with the threat of becoming disembedded from local culture, or overwriting local culture with broader corporate messaging and commercially driven narratives. This played out in a very public way at Surf Snowdonia in the lead up to the Red Bull Unleashed contest held at the facility 18–19 September 2015. The Welsh surfing community became increasingly angered by the absence of any local surfers on the contest roster (it is standard practice in high-level professional contests to include several local 'wildcard' entrants out of respect to the local surfing community). Even though Surf Snowdonia claimed to have no input on the make up of the list of contestants the international surf media ran with the headline 'Surf Snowdonia's PR Disaster in Wales: Why the Locals Are Angry' (Anderson, 2015). Eventually the contest was forced to acquiesce and add the Welsh surfing champion to the field of contenders (Inertia, 2015b), but not before significant damage had been done in the eyes of the Welsh surfing community upon whom the facility will be relying for a significant portion of their revenue. From a sustainability perspective it is important that local understandings of place are incorporated into the design and narratives of these kinds of recreation and tourism facilities.

An intensification of existing scepticism of surf park sustainability is an additional threat. Excessive and unsustainable use of land, water, and power were identified by 9 per cent of survey respondents as the reason they would not use a surf park. The following survey response is typical of this group: 'Surf pools seem like a colossal waste of land, fresh water and energy to create what the ocean gives us for free.' The survey revealed that respondents are also concerned about sustainability in their surf travel choices: 70 per cent say sustainability performance will influence their future surf travel choices and 91 per cent indicated they would choose a sustainable surf tourism business over non-sustainable if price and service offerings were comparable. Clearly perceptions of sustainability have the potential to significantly impact the market appeal of surf parks. While a range of myths and greenwashing distractions could be deployed by the surf park industry to head off the threats outlined above, real solutions to these threats are available.

Beyond Nirvanic myth

The opportunity exists for the surf park industry to transcend the status quo of myth making and reality maintenance by addressing threats head on with creative solutions that support the symbolic elements of its Nirvanic space, rather than applying the metaphoric Band Aid of obfuscatory rhetoric. For example, Surf Snowdonia is powered by an adjacent hydroelectric power plant that the development itself helped to refurbish and expand to service 20,000 homes in surrounding communities (Carrington, 2015). Additionally the water drawn to fill the lagoon is recycled from the outflow of the hydroelectric plant (Andy Ainscough, pers. comm., 2016). The land the project sits on was reclaimed and decontaminated after the closing of an aluminium factory. Kelly Slater Wave Co's test facility, which is built in a disused private water skiing lake, recently announced that it will be running on 100 per cent solar energy, though photovoltaic cells will not be erected onsite. While there is enormous variability between technologies and designs, a full-size pneumatic caisson surf pool at full power would draw in the neighbourhood of four megawatts of power. Four acres of photovoltaic cells are currently required to produce a megawatt of power, so creating enough power for a large pool would require around 16 acres of photovoltaic cells at current efficiency levels (John Luff, pers. comm., 2016). The Slater pool uses artesian water, which has attracted some negative media attention as an extravagance during California's worst recorded drought. This should, however, be put in context with other recreation facilities. A surf pool in southern California is likely to need refilling completely once per year due to evaporation, assuming no pool draining is required for repairs or maintenance.

This is the equivalent water usage of one hole on a golf course over the same period (John Luff, pers. comm., 2016). Different technologies have different energy and water requirements and differing implications if repairs are needed. For example, Surf Snowdonia was completely drained three times in its first two months of operation with 40 million litres of water required to refill on each occasion. Pneumatic caisson pools have the advantage of their wave generating machinery being accessible outside the pool.

Conclusion

Many of the technological, social, and cultural indicators reviewed here suggest that the surf park industry is on the verge of major development acceleration. This chapter has applied Nirvanification theory to explore sustainability issues related to the construction of wholly artificial surfing spaces as the industry progresses. The symbolic elements of Nirvanic surf park space were considered by surf park developers to be: perfect waves, guaranteed wave count, cushioned adventure, authenticity from a social/cultural perspective, the setting, and accessibility. Surf parks that can generate high-quality waves will have the Holy Grail effect – a reliably perfect wave with a wave count guarantee and the ability to manage the resource in a way that makes the most economic sense. However, these symbolic elements are threatened by the spiritual divide (the missing spiritual component cited by the vast majority of survey participants who would not use a surf pool), perceptions of surf pools being a waste of land, energy, and water, and the overriding of local cultural concerns and understandings of place. Rather than deploying Nirvanic myth to nihilate these threats, a more sustainable response from the industry would be to respond to them in ways which support the symbolic elements.

Bridging the spiritual divide in the surf pool experience may not be possible. Rather than attempting to usurp the ocean surfing experience, surf park surfing can be positioned as an experience that is complementary to ocean surfing and allows for focused creative expression in the act of surfing itself without the distractions of competing for waves and substandard surfing conditions. Land, power, and water usage issues can be incorporated into surf park narratives to find real, creative solutions (such as those utilized at Surf Snowdonia) and educate guests about the issues. Third party sustainability certifications specifically developed for Surf Parks, such as those developed for surf and snow resorts, may be useful tools for streamlining and auditing sustainability practices and for communicating these with surf park guests. The incorporation of local cultural concerns is vital to avoid the disembedding of surf pools from their geographic and cultural context. Interpreting key issues facing the world's oceans, and linking the surf park surfing experience to these issues at real surf breaks could help to forge such connections. Incorporating education about these issues, surfing history, and surf etiquette as part of surf park learning programmes could help to ground guests in the broader surf culture and lead to better experiences when/if they do surf in the ocean.

The surf park industry provides a wealth of future research opportunities that can help guide the industry to a sustainable trajectory. Further sociological explorations of the implications of simulating natural spaces can yield insights into key elements of the experience that maximize personal benefits to the participant despite the spiritual divide. Assessments of existing experiences and the efficacy of sustainability messaging could help to refine the way surf parks integrate sustainability into the surfing spaces and experiences they create. Tracking the economic and social impacts of surf parks on surrounding communities, and surrounding surfing communities will be very important to guide future developments. Analysis of visitation patterns for surf parks may yield new information about the demographic reach of this new kind of attraction, its potential reach as a tourist draw, and its distance decay characteristics. Can surf parks be legitimate international tourism destinations in the way that ocean surf breaks are? Will surf parks have the effect of reducing international surf tourism if high-quality, uncrowded surfing experiences are possible closer to home? If this industry is to be sustainable, it will require significant guidance from the academic and NGO communities.

References

AEC Group (2009). *Surf Industry Review and Economic Contributions Assessment*, Report prepared for the Gold Coast City Council, Gold Coast, Australia.

Anderson, T. (2015). Surf Snowdonia's PR disaster in Wales: Why the locals are angry. *The Inertia*. Retrieved 5 September 2015 from www.theinertia.com/surf/surf-snowdonias-pr-disaster-in-wales-why-the-locals-are-angry/.

Barbieri, C. and Sotomayor, S. (2013). Surf travel behavior and destination preferences: An application of the serious leisure inventory and measure. *Tourism Management, 35*(April), 111–121.

Bartholomew, W. (2015). The making and breaking of a surfer's paradise. In G. Borne and J. Ponting (eds), *Sustainable stoke: Transitions to sustainability in the surfing world*. Plymouth, UK: University of Plymouth Press.

Berger, P. L. and Luckmann, T. (1966). *The social construction of reality: A treatise in the sociology of knowledge*. New York: Doubleday.

Blunt, W. (1970). *The dream king: Ludwig II of Bavaria*. New York: Viking Press.

Budruk, M., Wilhem Stanis, S. A., Schneider, I. E., and Heisey, J. J. (2008). Crowding and experience use history: A study of the moderating effect of place attachment among water-based recreationists. *Environmental Management, 41*, 528–537.

Carrington, D. (2015). Wales factory site turned into world's longest man-made surfing wave. *Guardian*. Retrieved 5 June 2015 from www.theguardian.com/environment/2015/jun/05/from-aluminium-smelter-to-surfers-mecca.

City of Gold Coast (2016). *Gold Coast Surf Management Plan*. City of Gold Coast.

Coffman, M., and Burnett, K. (2009). *The value of a Wave: An analysis of the Mavericks Region, Half Moon Bay, California*. San Francisco, CA: Save the Waves Coalition.

Csikszentmihalyi, I. S. (1975). *Beyond boredom and anxiety: The experience of play in work and games*. Oxford: Jossey-Bass.

Cunningham, P. (2006). Social valuing for Ogasawara as a place and space among ethnic host. *Tourism Management, 27*, 505–516.

Davis, J. S. (2005). Representing place: 'Deserted isles' and the reproduction of Bikini Atoll. *Annals of the Association of American Geographers, 95*(3), 607–625.

Edensor, T. (1998). *Tourists at the Taj: Performance and meaning at a symbolic site*. London: Routledge.

Foucault, M. (1986). Of other spaces. *Diacritics, 16*(1), 22–27.

Giddens, A. (1990). *The consequences of modernity*. Stanford, CA: Stanford University Press.

Gold Cost City Council (2009). *Surf industry review and economic contributions assessment*. Gold Coast City Council.

Hopkins, J. (1998). Signs of the post-rural: Marketing myths of a symbolic countryside. *Geografiska Annaler, 80 B*(2), 65–81.

Inertia (2015a). 2020 Tokyo organizers recommend Olympic surfing takes place in the ocean, not wave pools. *The Inertia*. Retrieved 21 October 2015 from www.theinertia. com/surf/2020-tokyo-organizers-recommend-olympic-surfing-takes-place-in-the-ocean-not-wave-pools/.

Inertia (2015b). Intense public pressure adds local surfer to Red Bull Unleashed at Surf Snowdonia. *The Inertia*. Retrieved 9 September 2015 from www.theinertia.com/surf/ intense-public-pressure-adds-local-surfer-to-red-bull-unleashed-at-surf-snowdonia/.

Johnston, C. (1992). *What is a social value? A discussion paper*. Canberra: Australian Government Publishing Service.

Lollback, R. (2015). 'Surfable' wave pool would protect Ballina's surfers. *The Northern Star*. Retrieved 10 September 2015 from www.northernstar.com.au/news/surfable-wave-pool-would-protect-ballinas-surfers/2770391/.

Lowry, T. (1985). His money comes in Waves. *Sunday Call-Chronicle*. Retrieved 10 September 2015 from http://articles.mcall.com/1985-06-23/news/2465673_1_waves-pools-oil-rigs.

McElroy, N. (2016). Pumping surf adds $20 million to the Gold Coast economy in two months. *Gold Coast Bulletin*. Retrieved 1 March 2016 from www.goldcoastbulletin. com.au/lifestyle/beaches-and-fishing/pumping-surf-adds-20-million-to-the-gold-coast-economy-in-two-months/news-story/53d2ab7dbae9b5177fb07352d2cdc56f.

Manning, R. (2007). *Parks and carrying capacity: Commons without tragedy*. Washington, DC: Island Press.

Margules, T. (2011). *Understanding the roles of ecosystem services in the local economy of Uluwatu, Bali, Indonesia*. (BSc Hons), Southern Cross University.

Martin, S. A. and Assenov, I. (2012). The genesis of a new body of sport tourism literature: A systematic review of surf tourism research (1997–2011). *Journal of Sport & Tourism, 17*(4), 257–287. doi:10.1080/14775085.2013.766528.

Ministry of Tourism Arts and Culture. (2012). *Resort Boundary Regulation 2012/R-7*. Retrieved 4 March 2013 from: www.tourism.gov.mv/downloads/regulations/ Boundary_Regulation.pdf.

Murphy, M. and Bernal, M. (2008). *The impact of surfing on the local economy of Mundaka, Spain*. San Francisco, CA: Save the Waves Coalition.

Navarro Jurado, E., Mihaela Damian, I., and Fernandez-Morales, A. (2013). Carrying capacity model applied in coastal destinations. *Annals of Tourism Research, 43*, 1–19.

O'Brien, D. and Eddie, I. (2013). *Benchmarking global best practice: Innovation and leadership in surf city tourism and industry development*. Paper presented at the Global Surf Cities Conference, Kirra Community and Cultural Centre.

Pamer, M. (2015). Pro surfer Kelly Slater unveils 'best man-made wave ever' in secret location. *KTLA5*, 21 December.

Ponting, J. (2008). *Consuming Nirvana: An exploration of surfing tourist space.* (PhD (Leisure and Tourism)), University of Technology, Sydney.

Ponting, J. (2014). Comparing modes of surf tourism delivery in the Maldives. *Annals of Tourism Research, 46,* 163–165.

Ponting, J. (2015). *Constructing Nirvana: Implications for the surf park industry.* Paper presented at the Sustainable Stoke, San Diego State University.

Ponting, J. and Lovett, E. (2015). *Telo Islands sustainable surf tourism management plan.* Unpublished.

Ponting, J. and McDonald, M. (2013). Performance, agency and change in surfing tourist space. *Annals of Tourism Research, 43,* 415–434.

Ponting, J. and O'Brien, D. (2014a). Liberalizing Nirvana: An analysis of the consequences of common pool resource deregulation for the sustainability of Fiji's surf tourism industry. *Journal of Sustainable Tourism, 22*(3), 384–402. doi:10.1080/096695 82.2013.819879.

Ponting, J. and O'Brien, D. (2014b). Regulating Nirvana: Sustainable surf tourism in a climate of increasing regulation. *Sport Management Review, 18*(1), 99–110.

Preston-Whyte, R. (2001). Constructed leisure space: The seaside at Durban. *Annals of Tourism Research, 28*(3), 581–596.

Preston-Whyte, R. (2002). Constructions of surfing space at Durban, South Africa. *Tourism Geographies, 4*(3), 307–328.

Scorse, J., Reynolds, F., and Sackett, A. (2015). Impact of surf breaks on home prices in Santa Cruz, CA. *Tourism Economics, 21*(2), 409–418.

Temperley, E. (2015). The wave garden at Snowdonia will open in July. *Magic Seaweed.* Retrieved 23 January 2015 from http://magicseaweed.com/news/the-wavegarden-at-surf-snowdonia-will-open-in-july/7154/.

Tomson, S. (2013). *Waves of stone.* Paper presented at the Surf Park Summit, Costa Mesa.

Tuan, Y.-F. (1975). Rootedness versus sense of place. *Landscape, 24,* 3–8.

Vaske, J. J. and Shelby, L. B. (2008). Crowding as a descriptive indicator and an evaluative standard: Results from 30 years of research. *Leisure Sciences: An Interdisciplinary Journal, 30*(2), 111–126.

Wagner, G. S., Nelson, C., and Walker, M. (2011). *A Socioeconomic and Recreational Profile of Surfers in the United States: A Report by Surf-First and the Surfrider Foundation.* Surfrider Foundation.

Warshaw, M. (2004). *The Encyclopaedia of Surfing.* New York: Penguin Books.

Waveloch. (2016). Waveloch history. Retrieved 20 January 2016 from www.waveloch.com/history/.

Westwick, P. and Neushul, P. (2013). *The world in the curl: An unconventional history of surfing.* New York: Crown Publishers.

Wilson, J. (2008). That's bullshit. *Surfing Magazine.* Retrieved 20 January 2016 from www.surfingmagazine.com/news/thats-bullshit-chapter-1-jimmy-wilson-082208/ –640UZvkA12bw1P7P.97.

World Surf League (WSL) (2016). Sponsorship. Retrieved 20 January 2016 from www.worldsurfleague.com/pages/sponsorship.

Part VI

Conclusion

14 Sustainability and surfing

Themes and synergies

Gregory Borne

Introduction

In Part I and the introduction to this volume I drew together the narratives of sustainability and surfing couched within the broader discussions of a risk society. Sustainability was introduced as a contested and ambiguous concept and surfing was outlined as an activity that has no edges and is unique in its ability to explore issues of sustainability. The range of themes that currently exist within the surfing research field was presented and a selection of pertinent literature reviewed that further drew the relationship between risk, sustainability and surfing into focus. Throughout the book each author has contributed to the relationship between sustainability and surfing in their own way, building up a progressive picture of this relationship that has added depth and sophistication to initial observations. In this final chapter I will explore each contribution in turn, relate these to some broader debates on sustainability as well as suggest future directions for this exciting field of research.

Part II contains a single contribution from Steve Martin and Danny O'Brien. This chapter sets the context for the complex integration of the physical and the social worlds that fits very directly in line with the overall frame of this book. The authors build a progressive picture of the integrated nature of surf site boundaries that are interconnected beyond the specfic site of surfing exploring the physical features of the surfing habitat and the impact that humanity is having on the surfing zone. Focusing on surf habitat conservation this sphere of influence is then expanded to the interaction between humanity and the surfing zone that draws explicitly on the language of sustainability 'whereby individuals are concerned with using natural areas in ways that sustain them for current and future generations of human beings and other forms of life' (page 25). At this point the idea of stakeholder involvement and the development of management plans are also integrated into the sustainability scenarios where the historical and cultural components are also highlighted.

The contested and complex nature of this relationship is emphasised through the lens of the surfing reserve. This resonates with a number of authors in this volume who have underlined the importance of the surfing reserve which provides a unique opportunity to explore the relationship between surfing and sustainability from multiple perspectives. In particular, surfing reserves provide an opportunity to explore one of the central concerns of sustainability: of meeting complex needs and the pervasive relationship between economic development and preservation. The emphasis on surf resource stakeholders enables an explicit recognition of surfing's engagement with not only the coastal zone but also beyond this to include surfers and other members of the community that own or work in surf-related establishments, surf shops, surfboard makers and surfing schools. These stakeholders are expanded even further when considering surfing events. The chapter suggests that 'published research attesting to the physical and human "surf system" as a holistic spectrum of social, economic, and environmental criteria and implications for sustainability is limited' (page 30).

The opening up of stakeholder networks has been discussed in the broader sustainability and sustainable development literature largely within the context of governance (Lafferty 2004). The Commission on Global Governance defines the term as:

> The sum of the many ways individuals and institutions, public and private, manage their common affairs, a continuing process through which conflicting and diverse interests may be accommodated and co-operative action may be taken.
>
> (Shridath and Carlsson 1998:2)

Within this context the authors argue that two paradigms exist when looking at surfing sites in the social sciences. The first is the global value perspective (industry) and the second, the values embedded from local communities and individuals. Moreover, a whole systems approach has the following advantages. First, it revolutionises the understanding of coastal systems, community and sustainability. Second, it augments the role of social and environmental sciences. Third, it expands the epistemological base including multidisciplinary and mixed methods research. Martin and O'Brien's contribution is representative of broader debates within the sustainability literature on complexity and the need to explore sustainability-related issues through a complex systems approach in order to obtain a realistic understanding of the issues involved (Harris 2007; Grin *et al.* 2010).

Part III continues to explore the complexity of the relationship between sustainability and surfing, applying a critical lens to technological advancement in different contexts. The opening chapter by Leon Mach skilfully integrates a broad-ranging discussion on the role of technology, surfing and sustainability. Outlining a critical discussion of technological determinism the chapter highlights the advantages of using a technology, environment and society framework. This is then applied to what the author describes as the four technological dimensions that include, physical, climatology, Internet communication technology and artificial surfing. This chapter provides not only a broad-ranging discussion but also explores the epochal transition from a technological era to one of sustainability. These discussions resonate strongly with my exploration of modernity as understood in a risk society. This reinforces the strength of surfing as a subject that can critically explore these meta level theoretical discussions. For Mach the era of sustainability is underpinned with an emphasis on agency and the ability to steer surfing onto a more sustainable path. This, Mach argues, is demonstrated through a series of conferences and events that have emerged in the past three years.

The second chapter in this section focuses on the role of clusters in achieving sustainability in the surfing industry. Anna Gerke explores the development of the French surf industry in the Aquitaine regions and the impacts on the local industry dynamic. This is an exciting addition to the literature both from a geographical perspective and as a perspective on industry development. Gerke introduces the term of clusters in the facilitation of sustainable surfing. Interestingly, Gerke also stresses the relationships between surfers that form bonds that are stronger than that to a single company. This is not taken as a negative but actually integrated as a positive attribute. This is an interesting finding and as Gerke points out: 'The joint practice of surfing while being sometimes employed by direct competitors creates social links between employees across different cluster organisations via the sport' (page 79).

Gerke identifies some of the opportunities and barriers that exist for achieving sustainable business and emphasises the role of overarching bodies, in this case EuroSIMA, to help coordinate and promote sustainability within the clusters where a competitive and profit-orientated perspective dominates. This chapter also highlights tensions that can exist at the global and local levels and echoes broader observations relating to the development of the surfing industry and its continued expansion (Laderman 2015). Gerke also emphasises the importance of the difference between the legal and discretionary dimensions of integrating sustainability into operating procedures and the overall culture of the companies.

Chris Gibson and Andrew Warren explore the pivotal role of surfboards in surf culture and what this means for sustainability. Whilst a number of authors in the book allude to the role of surfboards within surfing and particularly with a focus on the sustainability of materials, Gibson and Warren reinforce this with a discussion on the interaction between the global and local dynamics and informal subcultural scenes. Whilst emphasising the environmental aspect of sustainability the authors also explore the complexity of multiple sustainability-related issues.

Resonating with observations by Gerke, the chapter highlights the industry's local ties and points out that the soulfulness of surfing is still embedded in the manufacture of boards. However, despite multiple innovations the authors argue that 'inhibiting sustainability improvements are factors linked to the industry's informal DIY origins, which has given rise to a distinctive – and limiting – mix of economic structure and subcultural norms' (page 97). The infamous Blank Monday incident, when Clark Foam closed on the 5 December 2005, was one of the pivotal moments that altered the surfboard industry. This was precipitated by specific environmental and workplace safety issues. The authors explore the relationship between locally made boards and the rapid expansion of distribution globally, and the technological advancements through computer and CAD (CNC) systems such as Firewire and the impact this has on distribution and sustainability of surfboard production and consumption (see Hyman 2015). With that said, the authors indicate that local shapers in small workshops often still exist because of local demand at key surfing spots. Emphasising the role of the ECOBOARD project initiated by Sustainable Surf, the authors explore the evolution of core materials from polyurethane to the emergence of more ecologically sound materials and production techniques.

Part IV moved the debates of sustainability and surfing forward with an emphasis on activism, political advocacy and the relationship of various actors or knowledge exchange and the creation of effective policies (see Cvitanovic *et al.* 2015). Works that have explored these issues in surfing include Wheaton (2007), Thorpe and Rinehart (2013), and Laderman (2014, 2015). The clustering of authors in this section offers fascinating insights into an emerging relationship between sustainability and surfing. It develops a number of themes that are consistent throughout the other chapters and takes them forward through broader empirical work. Ware *et al.* and Hales *et al.* use the case study of Australia's Gold Coast to explore different areas of policy and activism. For Ware *et al.* 'Surfers have a long and proud history of protest and advocacy on the issues that reflect the connection of surfers to the coastal marine environment' (page 107). The emphasis in this work is on the lack of a surfing contract which creates the tension between commercial and environmental interests. This is applied to the agenda-setting process which has led to the establishment of the Gold Coast City Surf Management Plan. As the authors note:

> In contrast to many local government, management planning processes, the plan was instigated by a coalition of community surfing organisations and interested individuals as a mechanism to support the transition to a more institutionally recognised and socially acceptable social contract for surfing on the Gold Coast.
>
> (page 108)

The authors use as their foundation the multiple streams theory that focuses on the three streams of politics, problem recognition, and the formation and refining of the policy proposals, emphasising that these streams often operate independently but occasionally converge either by chance or by design. What is also unique about this chapter is that the authors acknowledge their role in the policy landscape, forming a participant observation methodological structure. Lead author Dan Ware was 'instrumental, along with others from the surfing organisations, in forming the participatory, ground-up approach to the Gold Coast surfing social contract policy problem by drawing on the multiple streams theory' (page 120).

Drawing on a range of data sources the authors effectively weave a narrative around each stream using highly relevant and topical examples. The authors highlight the eventual establishment of the surf management plan.

The previous adversarial relationship between surfers and government had negative consequences for the interests of both parties and the Surf Management Plan is a way to decrease these consequences.

(page 122)

This is again a consistent theme throughout this book as well as the broader debates that punctuate sustainable development and sustainability more generally. The discussion in the introduction relating to theoretical propositions of a reflexive modernity and ecological modernisation gives context to these debates. Drawing on the language of the risk society this represents the opening up of the political space through the idea of sub-politics. Whilst there are a number of articulations of what this entails in purer and more diluted forms, it is essentially the opening up and reconfiguration of the political processes. As I have already discussed, a central tenet of a reflexive modernity is humanist relationship with nature that has moved past the modernistic attitude of domination one of protection. And for Wheaton this is an attitude 'that is prevalent in surfing discourse' (Wheaton 2007: 283).

In the second chapter in this section Hales *et al.* extend the previous discussion by exploring the role that surfing protest has in coalescing the surfing community. In particular, the chapter explores the importance of the public sphere responses. As the authors point out: 'The public sphere is not a singular place but rather there are multiple spaces where the public can variously attempt to express matters that they consider important to society' (page 126). It is in these spaces that both public perception as well as political decision making can be influenced. The authors explore the hybrid notions of the space of the waves and the influence of the media on these spaces. These debates resonate with a number of commentators and social theorists who discuss the importance of the media in relation to environmental movements and society more broadly (Anderson 1997). For Beck the space of politics is not the street but the television and refers to this as real virtuality. This also resonates strongly with the role of discourse in political contexts. For Hajer (1995) discourses play a constituent role in policy making where policy making is seen as the struggle between various discourse coalitions though which storylines are kept together. Grounding his assertion in Foucault, Hager argues: 'Discourse are not to be seen as a medium through which individuals can manipulate the world as conventional social scientists suggest, it is instead part of reality and constitutes the discoursing subject' (1995: 61). This is important because it is recognised that the discourses that relate to sustainable development and sustainability exist within a discursive power play and cannot be divorced from overt rhetoric.

The authors' argument is further developed through the notion of enclosure where previously public good is included in the economic calculations and no longer offering value to society. This is applied to a typology of surfing protest that is linked to Lazarow's (2007) notion of wave capital highlighting both historical and contemporary protests globally. Following this the chapter focuses on a single case study on the Gold Coast to explore these dynamics in more depth. The authors explore the relationship between surfers' organising capacity, the increasing recognition of the economic importance of waves as well as the importance of surfing culture and appeal particularly in relation to the power of surfing celebrity to engender support. There are then multiple issues that converge in this chapter that add to the dynamic of surfing and sustainability.

Central to the discussion in Hale *et al.* is the idea of enclosure, the cordoning off of the public sphere land to reduce or negate the benefit to society more broadly. As the authors explain, identifying the economic value that translates into the policy makers' rubric of understanding played an important role in protest organisation and communication of the value of waves. Jason Scorse and Trent Hodges step outside the rubric of economistic analysis to explore the non-market value of surfing. The authors identify the growing work around surfonomics and points to Neil Lazarow's foundational work in the area indicating that:

> To date no study has yet been published on the non-use value of surfing. Such research would include the existence value people hold from simply knowing that surfing resources are being protected (even if they don't use them directly), the option value of preserving a surf location for future use, or the bequest value of being able to pass surfing resources down to future generations.
>
> (page 138)

The chapter identifies the importance of the use of a contingent value methodology which asked survey respondents to respond to hypothetical future situations. This is inline with a growing body of work on the nature of scenarios in achieving sustainability especially in response to perceived risk (Renn 2008; Hofmeester *et al.* 2012). Scorse and Hodges highlight the Travel Cost Method as the most common type of non-market value and that this has been used as a form of surfonomics. Drawing on the discussion in the introduction this suggests a mechanism that may be representative of a reflexive modernity as it is a departure from the modernistic, ecological modernity interpretation of how value is calculated within society. This also sits within a broader literature that explores ecological or sustainability economics (Soderbaum 2008).

Part V presents chapters that in some form explore the reconceptualisation of surfing spaces. Opening the discussion Lindsay Usher explores the relationship between localism and sustainability. Initially this chapter provides a very important outline of sustainability that in line with the introduction to this book emphasises ambiguity and the contextual nature of the concept. The chapter then proceeds to outline the nature of localism from extreme to mild localism. This encompasses multiple issues from surf rage to place and attachment. However, Usher points out that 'localism can also make positive contributions to the environmental sustainability of surfing' (page 153). Outlining a broad range of literature the chapter very effectively builds a discussion that focuses on the space of the waves in the context of sustainability. Reinforcing this debate the chapter presents the results of ethnographic work in both Nicaragua and Costa Rica concluding:

> While localism may challenge the economic sustainability of surfing, sustaining the local surf culture and local surfing space is important for surf communities. Positive forms of localism should outshine this phenomenon which is known as the dark side of surfing. By encouraging positive manifestations of localism and discouraging the negative ones, surfers can work towards a more sustainable form of surfing.
>
> (page 162)

The next chapter, by Mark Orams, extends the discussion presented in the previous chapter and explores the rise of surfing and overcrowding which can lead to localism introducing the idea of Spot X, where locations are not disclosed. The chapter delivers a very concise overview of a number of tourism and development models summarising:

> The irony of the focus on exploring and discovering new surfing destinations is that in doing so these explorers can unintentionally become the genesis of the very experience and outcome they detest and are seeking to escape.
>
> (page 169–170)

Orams' introduces the notion of wilderness where there is an overriding need to protect this space in the face of overwhelming overpopulation and the increased use of the surfing space. In the following chapter in this section Jon Anderson explores the place of the tube and the practice of surf travel. This chapter encourages the reader to seriously reflect on what global surf culture actually means. The tension in the chapter is whether surfing as an internally pleasurable event or whether this must be contextualised as rituals with external consequences. Anderson emphasises the feeling of surfing the sublime, the flow and of course the stoke. But with the stoke comes the expansion of a multibillion dollar industry and the externalities it produces. 'It is questionable whether the external consequences of these identity practices are reflected on in depth within this culture, and it is with this problematic that the language and practice of sustainability must engage' (page 178).

Anderson asks two very pertinent questions. First, what does sustainability have to do with a surfing utopia and second, why would surfing culture be willing to engage with this unwieldy concept more than any other group? These questions are fundamental to the exploration of sustainability in surfing. The chapter presents data on the carbon associated with surfing travel. The individualism within surfing is highlighted where it is argued that this individualistic instinct is not necessarily at odds with sustainability. 'Sustainability is increasingly framed in terms of the individual and the isolated citizen is now often seen as a key contributor to the environmental crisis and any future' (page 192). Furthermore,

> If environmental destruction, global warming and sea level rise were clearly understood as direct threats to the existence of certain breaks and waves, then becoming part of a movement that adopts more sustainable practices may make sense to a hedonistic surfer.
>
> (page 193)

These observations resonate strongly with multiple authors in this book, from discussion on activism and policy debates through to localism. Moreover, these debates on the nature of the individual are also a central theme within Beck's understanding of late modern structures. I have previously suggested that 'the process of individualisation as Beck understands them can inform sustainable development discourse by understanding how people compose their own identities and biographies in the face of increased globalised knowledge' (Borne 2010: 276). What Beck tells us about the process of individualisation in a reflexive modern world is that it occurs through the phases of liberation, detachment and reintegration. In this sense individualisation does not mean, 'atomisation, isolation, loneliness, the end of all kinds of society of unconnectedness' (1994: 13). These observations then compliment and support Anderson's comments by reformulating the notion of individualisation so that 'surfers may be able to extend their commitment to the transience of stoke, re-orientating their "liquid lives"' (page 198)

Extending the discussion, the next chapter in this section explores culture, meaning and sustainability and holds strong synergies with the previous chapter. Lazarow and Olive's discussion 'contributes to the growing number of voices critiquing established understandings of surfing history and hierarchies as well as the powerful role traditional media and industry have had in developing surfing's cultural identity' (page 204). This resonates strongly with this volume's introductory observations on historical narratives in a postmodern era. The authors explore the notion of attachment to place and the role that this plays in creating action in a particular locale, highlighting the importance of personal relevance in these discussions. Epistemologically the research presented draws on feminist cultural studies, which explores multiple themes through the lens of embodied subjectivities. Again resonating with other authors in the volume, there is an understanding of sustainability, 'in more holistic terms, as relating to surfing culture, localism, relationships (to place and people) and coastal / surf break management' (page 204). Indeed, in line with Martin and O'Brien, the authors argue that, 'the social and cultural world is intimately connected with the natural world and must be considered a "surfing system"' (page 214).

It is fitting that the final chapter in this section and indeed in the book explores the potential impact on sustainability and surfing from the increasing prevalence of artificial surfing spaces. This is an increasingly relevant debate for sustainability and surfing on multiple levels. The chapter provides a concise history and overview of surf parks before exploring the broader political and social landscapes within which they sit. Ponting's theory of Nirvanification is effectively applied to these emerging surfing zones. Through this lens surf pools are described as enclavic spaces. 'Under the guise of freedom and escape, these kinds of spaces are subject to intense social control, leaving little room for local understandings of space' (page 227). The chapter then moves on to explore some primary data, both survey and interview.

The chapter points out that those enclavic qualities need to be integrated into the design of the surf parks, manufactured in a controlled environment. And, interestingly these elements are balanced with a need to reduce risk perception of those fearful of the ocean. This creates an additional dimension to the relationship with the wave itself and what this means for surfing's connection to nature through the foundational oceanic experience. Drawing on the introductory observations relating to stoke and affect, this takes the discussion into a new realm of artificiality and complicates the relationship with the body and perception and behaviours towards the environment. Ponting outlines a number of future areas of research that this emerging technology can offer for the relationship between sustainability and surfing from the implications of simulating natural spaces, the efficacy of sustainability messaging through to the impact on local communities.

Overall this book has provided an initial foundation on which future research on sustainability and surfing can be taken forward, which includes theory, methodology, empirical evidence, practical application, and the relationship between these elements. As we have seen, each author has discussed a different aspect of surfing and contextualised this within specific surfing literature that draws together multiple components of sustainability and surfing. This in turn has been related to some of the broader debates that underpin sustainability discourse. Drawing finally on the introduction that explored sustainability and surfing from within the context of a risk society I make one final observation.

Ulrich Beck's final work, *Metamorphosis*, represents an extension of the risk society thesis and centres on the idea of emancipatory catastrophism. In this new work 'metamorphosis is not social change, not transformation, not evolution, not revolution and not crisis. It is a mode of changing human existence. It signifies the age of side effects' (Beck 2016: 20). Focusing on climate change the theory of metamorphosis suggests that 'the literature relating to climate change has become a supermarket for apocalyptic scenarios. Instead the focus should be on what is now emerging – future structures, norms and new beginnings' (Beck 2016: 39). I see *Sustainable Surfing* as part of the process of new beginnings but looking beyond the imagined perfection of the surfing ideal to new futures in an era of late modernity.

References

Anderson, A. (1997) *Media, Culture and the Environment*, London, Routledge.

Beck, U. (1994) *Ecological Enlightenment*, Atlantic Highlands, NJ, Humanities Press.

Beck, U. (2016) *The Metamorphosis of the World*, Malden, MA, Polity Press.

Borne, G. (2010) *A Framework for Sustainable Global Development and the Effective Governance of Risk*, New York, Edwin Mellen Press.

Cvitanovic, C., Hobday, A., Kerkhoff, K., Dobbs, K., and Marshall, N. (2015) Improving Knowledge Exchange among Scientists and Decision-makers to Facilitate the Adaptive Governance of Marine Resources: A Review of Knowledge and Research Needs. *Ocean and Coastal Management* 112: 25–35.

Grin, G., Rotmans, J., and Schot, J. (2010) *Transitions to Sustainable Development: New Directions in the Study of Long Term Transformative Change*, London, Routledge.

Hajer, M.A. (1995) *The Politics of Environmental Discourse: Ecological Modernization and the Policy Process*, Oxford, Clarendon Press.

Harris, G. (2007) *Seeking Sustainability in an Age of Complexity*. Cambridge, UK, Cambridge University Press.

Hofmeester, C., Bishop, B., Stocker, L., and Syme, G. (2012) Social Cultural Influences on Current and Future Coastal Governance. *Futures*, 44(8): 719–729.

Hyman, N. (2015) Sustainable Transitions: From FireWire to Nevhouse. In G. Borne and J. Ponting (eds) *Sustainable Stoke: Transitions to Sustainability in the Surfing World*, Plymouth, UK, University of Plymouth Press, pp. 112–117.

Laderman, S. (2014). *Empire in Waves: A Political History of Surfing* (Vol. 1). Berkeley, CA, University of California Press.

Laderman, S. (2015) Beyond Green: Sustainability, Freedom and Labour of the Surf Industry. In G. Borne and J. Ponting (eds), *Sustainable Stoke: Transitions to Sustainability in the Surfing World*, Plymouth, UK, University of Plymouth Press, pp. 80–83.

Lafferty, W. (ed.) (2004) *Governance for Sustainable Development: The Challenge of Adapting Form to Function*, Cheltenham, UK, Edward Elgar.

Lazarow, N. (2007) The Value of Coastal Recreational Resources: A Case Study Approach to Examine the Value of Recreational Surfing to Specific Locales. *Journal of Coastal Research* (S1)50: 12–20.

Renn, O. (2008) *Risk Governance: Coping with Uncertainty in a Complex World*, London, Earthscan.

Shridath, R. and Carlsson, I. (1998) *Our Global Neighbourhood: The Report of the Commission on Global Governance*, Oxford, Oxford University Press.

Soderbaum, P. (2008) *Understanding Sustainability Economics*, London, Earthscan.

Thorpe, H. and Rinehart, R. (2013) Action Sport NGOs in a Neo-liberal Context: The Cases of Skateistan and Surf Aid International. *Journal of Sport & Social Issues* 37(2): 115–141.

Wheaton, B. (2007) Identity, Politics, and the Beach: Environmental Activism in Surfers Against Sewage. *Leisure Studies* 26(3): 279–302.

Index

Page numbers in *italics* denote tables, those in **bold** denote figures.

Abel, A. 34
accessibility, surf parks 230
acetone 91, 93, 94, 95, 96, 100
affect 4, 10, *11*, 178
aggression 148, 150, 153, 156, 168, 171, **212**, 213
Aguerre, Fernando 224
Aikau, Eddie 149
Ainscough, Andy 227, 229–30, 231, 232
air transport 54, 176–7, 181–2, *183*, 196
Alder, J. 28
algae 63, 95
Alter, Hobie 90
American Wave Machines 221, 222
Anderson, A. 4
Anderson, Jon 148
Aquaboggan water park, Saco, Maine 220
Aquitaine surf industry cluster 72–3, **74**, *75*, *76*, 243; collective sustainable practices 81–3; firm-level sustainable practices 80–1; location-specific factors 76–80
Aramoana, Dunedin, New Zealand *129*
Arctic Foam 95
Aristotle 44
artificial reefs 24, 118, 119
artificial surfing **51**, 59–60, 66–7; *see also* surf pools and parks
artificial surfing reefs (ASRs) *11*, 60, 66–7, 151, 173
Asbury Park, New Jersey *131*
Asia 91–2; *see also* Indonesia
ASRs *see* artificial surfing reefs (ASRs)
Assenov, I. 13, 30, 178
Association of Surfing Professionals 223
asthma 93, 95, 100
Asturia, Spain *129*

attachment to place 205–6
Augustin, J. P. 32
Australia: artificial surfing reefs 60; campaigns 114–17, *129*, 131–3, *132*; economic value of surfing 29, 137, 222–3; employment 29; government elections 133; localism 156–7; paulownia cultivation 97; shark attacks 214, 223; surf industry clusters 78; surfboard manufacturing 90, 97, 98, 100, 101; surfing reserves 27, 29, 115–17; tourism development 167–8; *see also* Gold Coast Surf Management Plan
Azores, Portugal *129*

Bahia de Todos Santos 137
Baker, S. 7
Baker, Tim 202
Bali, Indonesia 61, 152, 157, 167–8, 180–5, **181**, **182**, **184**, 189, 190
Ballina, Australia 223
balsa (*Ochroma pyramidale*) 96–7
bamboo 95
Banatao, Desi 95
Banatao, Rey 95
Barilotti, S. 180, 183
Barr and Wray 220
Bartholomew, Wayne Rabbit 114, 119, 149, 223
beaches 24, 27; campaigns to protect 128, 130, *130*, *131*, 153; *see also* surf sites
Beaumont, E. 155
Beck, Ulrich 6, 7, 8, 245, 248, 249
Bells Beach, Victoria, Australia 27
BenPat International 92
Bernal, M. 31, 137

Betts, Greg 118, 120
Biewer, Frank 220
Big Surf pool, Tempe, Arizona 220
biomechanics *11*
biophysical risks 210–11, **212**, 213
bioresins 63, 80, 95, 101
Black, Kerry 119
Blank Monday 88–9, 93, 94, 95
blanks, surfboard 52; environmental issues
  93–4; environmentally friendly
  materials 62, 63, 80–1, 95, 96, 99, 101;
  glassing 91, 92, 94–5, 99–101; health
  and safety issues 94; regulation 88–9,
  93–4, 95, 100; shaping 90, 91–2, 94, 98,
  100–1
Booth, Doug 4, 10, 12
Borne, G. 61
Boscombe Reef, England 151
Bourez, Michel 101
Bra Boys 156–7
Brazil 60
Brigham Young University-Hawai'i 138
'Bring Back Kirra' campaign 114–15, 132
Broadbench, England *131*
Broadhurst, R. 28
Broadwater Marine Project, Gold Coast,
  Australia 110
Brown, D. 4, 14, 155, 160, 205, 206
Bruntland Commission 4–5
Buckley, R. C. 13, 25, 32, 34
Burnett, K. 140
Butler, Richard 167–8, 169
Butt, T. 29, 30, 31

California 24, 53; campaigns *129, 130,
  131*, 153; economic value of surfing
  137, 139–40, 141, 223; localism 149,
  152, 156; surfboard manufacturing
  88–9, 95, 96, 97, 98, 100; tourism
  development 167–8
cameras 47, 54–5, 172
campaigns 66, 107–8, 114–17, 125–34,
  129, 130, 131, 132, 153, 205, 245–6
cancers 94, 100
caoutchouc 82
capitalized real estate values 138, 140–2,
  223
carbon emissions 8, 93, 96
carbon footprints 181–2, *183*
Carbon Gym software 181, 182, *183*
carbon offsetting 196–7
Carlsson, I. 242
Carson, Rachel 5
Casey, S. 177

cash localism 148
castor oil 96
Center for Surf Research 61, 173, 224
Centre for Alternative Technology 181,
  183
Chapman, D. J. 139
China 91, 92
Clark, Gordon 'Grubby' 88–9, 90
Clark Foam 52, 88–9, 90, 93, 94
climate change 4, 8, 223
climatology **51**, 56–7, 63–4
cluster theory 78
clusters *see* Aquitaine surf industry cluster
coastal conservation 25–6, 27–9, 153, 170,
  174
coastal development 165; campaigns
  against 66, 107–8, 115–17, 125–34, *129,
  130, 131, 132*
coastal management *11*, 107–9, *109; see
  also* Gold Coast Surf Management Plan
Coffman, M. 140
Cohen, M. 109
Colas, A. 177
Collins, Sean 56
colonialism: cultural 53, 184–5, 186–91,
  192–5; and localism 149–50, 157
Comley, C. 156, 158
commercialisation of surfing 150
Commission on Global Governance 242
commodification of surfing 13–15, 41
Commoner, Barry 5
community development 33
competitions *see* surfing events and
  contests
computerized shaping 90, 91, 92
consequences of surfing travel 176–98,
  247–8; carbon footprints 181–2, *183*;
  crowding 151, 155, 169, 182–4; cultural
  colonialism 53, 184–5, 186–91, **192–5**;
  on developing countries 13, 32–3, 34,
  53–4, 180–9, **181, 182, 184–95**;
  environmental impacts 53–4, 181–2,
  183, 185–6, **187–91**; and the individual
  surfer 191–6; internal 178; politics of
  pragmatism 196–7; *see also* localism
conservation 25–6, 27–9, 153, 170, 174
Conservative Party, Australia 133
consumer surplus values 138, 139–40
contests *see* surfing events and contests
Contingent Valuation Method (CVM) 138
Convention on International Trade in
  Endangered Species (CITES) 97
Corasanti, N. 59
Cornwall, England 130, 167–8

corporate social responsibility 72
Costa Rica *130*, 155, 158–61
Costco 52, 92
countercultural epoch 41
countercultural movement 15, 125
coyote environmentalism 197
creative destruction 45
critical epochs 48
crowding: and artificial surfing 67, 223, 225, 231; and exclusive access at surf resorts 224–5; and localism 150–2, 160–1, 165–6, 171; and safety 150–2, 166, 167; and surf etiquette 148, 151–2, 157, 166; and surf management 107, 111–12, 113, 115, 117; and surf rage 148, 150, 168; and surf tourism/travel 151, 155, 169, 182–4
cruise ship terminals, Gold Coast, Australia 110–11, 116, *129*, 131–2, *132*
Csikszentmihalyi, I. S. 178, 228
cultural colonialism 53, 184–5, 186–91, **192–5**
Currumbin Alley, Gold Coast, Australia 112–13, *132*
CVM *see* Contingent Valuation Method (CVM)

Dana Point, California *129*
Daskalos, C. T. 149
Deane, Wayne 112, 114
demarcation of surf sites 26–7, 149
destination life-cycle model 167–8, 169
determinism 45; technological 45–7
developing countries: impacts of surf tourism/travel 13, 32–3, 34, 53–4, 180–9, **181**, **182**, **184–95**; localism 156, 157, 158–61; property taxes 142; surfboard manufacturing 52; traditional resource custodians 33–5
digital public opinion 66
disabilities, surfers with 153–4
Disney World Orlando, Florida 220, 221
displacement of local communities 169
diversity 215
Dominican Republic *131*
Dorney Park Wildwater Kingdom, Allentown, Pennsylvania 220
Dorset, W. 149, 157
Doxey, G. V. 168, 169
drones 47, 54, 172

Eckerberg, K. 7
Ecoblanks 96

ECOBOARD certification scheme 63, 95–6, 99, 101
EcoComp UV-L resin 96
ecological modernisation 7
ecologically connected imagery 43–4
economics of surfing *11*, 14, 222–3; direct market values 137–8; and localism 155–6; non-market values 137–42, 223, 246; stakeholders 29–30; surfboard manufacturing 97–101
ecosystem-based management 28
Ecuador 96–7
Edensor, T. 225–6
education 32, 108, *109*
El Salvador 155
elections 133
Ellul, Jaques 45, 47
Empire Pool, Wembley, London 220
employment 29
enclavic spaces 225, 227
enclosure of the commons 127–8
England: artificial surfing reefs 60, 151; campaigns 129–30, *130*, *131*, 153; Global Wave Conference 61; localism 157; surf parks 229; surf pools 220, 222; surfboard manufacturing 90, 96; tourism development 167–8; *see also* Wales
entrepreneurship empowerment 33
Entropy Boards 95
Envirofoam blanks 96, 99
environmental degradation 8, 25–6, 53
environmental health *see* occupational health and safety
Environmental Impact Assessments 26
environmental issues 108; protests linked to 128, 130; surf parks 233–4; surf tourism/travel 53–4, 181–2, *183*, 185–6, **187–91**; surfboard manufacturing 93–4; *see also* environmental degradation
Environmental Protection Agency (EPA) 100
environmentally friendly materials 62, 63, 80–1, 95–7, 99, 101
epistemological Luddism 64–5
epoxy resins 80, 91, 93, 95
EPS *see* expanded polystyrene (EPS)
equipment 47, 50–3, 172; sustainable practices 61–3, 80–3, 95–7, 99; *see also* surfboard manufacturing
erosion mitigation 119
ethics 32; *see also* surf etiquette
etiquette, surf 148, 151–2, 154, 157, 166
European Association of Surfing Doctors 174

EuroSIMA 73, 75, 77, 78, 81, 82, 83
events *see* surfing events and contests
Evers, Clifton 179, 195
expanded polystyrene (EPS) 90, 91, 93,
    101; recycled 62, 63, 80, 96, 99
extended self model 195–6
extruded polystyrene (XPS) 90, 91

Facebook 126, 132, 176, 219
Fanning, Mick 110, 116, 133, 223
Farmer, B. 27, 28, 29
female participation 215
fibreglass cloth laminates 91, 93, 95, 99
Fight Club (film) 47
Fiji 35, 127, 224–5
fill coats 91
Finnegan, W. 204
Firewire 63, 101
floodlighting surf breaks 113
flow 178, 228
Flow Barrel, Norway 220
Flowrider, Texas 220
foam/fibreglass surfboards 51–2, 87, 88–9,
    90–2, 93
Ford, N. 4, 14, 160, 205, 206
Fordham, M. 183
forecasting, surf 56–7, 58, 63–4, 172, 176,
    225
forestry management 96–7
formaldehyde 94, 100
Foucault, Michel 226, 245
foundational era 41, 48–9
France *see* Aquitaine surf industry cluster
Frazer, R. 157
Freshwater Beach, Australia 26

Garajagan, Java, Indonesia 224
garbage can model of organisational
    choice 109
Gellert Baths, Budapest 220
genetic technology 8
Germany 220
ghost shaping arrangements 98
Gibson, C. 12, 88–9
Gili Eco Trust 185
Gili Trewangen, Indonesia 185–9, **185–95**
glassing 91, 92, 94–5, 99–101
Gliss Expo Seignosse 75
Global Surf Industries 92
Global Wave Conference 61
Gold Coast, Australia 32; campaigns to
    protect 114–17, *129*, 131–3, *132*;
    economic value of surfing 29, 137,
    222–3; surf industry cluster 78; surfing

reserves 115–17; *see also* Gold Coast
    Surf Management Plan
Gold Coast National Surfing Reserve 115
Gold Coast Surf Council (GCSC) 121
Gold Coast Surf Management Plan
    107–22, 223, 244–5; development of
    plan 120–2; policy proposals 109,
    113–17; politics 109, 118–19, 120;
    problem recognition 109, 110–13
Gold Coast World Surfing Reserve 115–17
Google Earth 172
government elections 133
grassroots surf organisations 32
green wash 5
Groundswell Society 202, 211–12

habitat, surfing *see* surfing habitat
Hajer, M.A. 245
Hamilton, Laird 177
Hanneman, W. M. 139
Hannigan, J. 7
Haraway, D. 197
Hawai'i, ancient 48–9; demarcation of surf
    sites 26–7, 149; inception of surfing 41,
    48; Kapu system 48–9, 149; surfboard
    manufacturing 48–9, 87
Hawai'i, modern: campaigns *130*;
    economic value of surfing 138; localism
    149–50, 152, 155–6, 158; surfboard
    manufacturing 97, 101; surfing events
    and contests 59; tourism development
    167–8; wave sharing 166
health issues, surfboard manufacturing 51,
    83, 94–5, 99–101
Hedonic Price Method (HPM) 140, 223
hemp cloth laminates 95, 96, 99
Hening, Glenn 211–12
Hess Surfboards 63
heterogeneous spaces 225–7
history of surfing *11*, 12–13, 41, 48–50
Höglander, Hofrat 220
holy grail effect 224–5
hot coats 91
Hounsfield, Nick 227, 229
Houses of Parliament, London 61
HPM *see* Hedonic Price Method (HPM)
Huanchaco, Peru 137
Huff, Lenard 138
Hui O He'e Nalu 152
hydroelectric power 233

Iatarola, B. M. 54
ICTs *see* Internet communication
    technologies (ICTs)

image capturing technologies 47, 54–5, 172

imagery, surf 43–4, 54–5, 58–9

India 60

indigenous communities: cultural colonialism 53, 184–5, 186–91, **192–5**; traditional resource custodians 33–5; *see also* Hawai'i, ancient; local communities

Indonesia 33, 59, 91, 180–9, *181*, **182**, **184–95**, 224, 225; *see also* Bali, Indonesia

Ingersoll, K. E. 149

International Hygiene Exposition, Dresden 220

International Surfing Association 155, 222, 224

Internet communication technologies (ICTs) **51**, 57–9; peer-to-peer information sharing 58–9, 65–6; surf forecasting 56–7, 58, 63–4, 172, 176, 225; surf imagery 54–5, 58–9; surrounding gaze politics 66

Irritation Index (Irridex) 168, 169

Irwin, J. 205

Ishiwata, E. 158

IUCN Red List of Threatened Species 97

Jacobs, M. 179

Japan 220

Jeffrey's Bay, South Africa *130*, 167–8

Jessen, S. 28

Johnson, J. C. 206

Kaczynski, Ted 47

Kaffine, D. T. 156

Kapu system, Hawai'i 48–9, 149

Kay, R. 28

Kelly Slater Wave Company 219, 221, 222, 224, 227, 233

Kildow, J. 139

Kingdon, J.W. 109

Kirra cruise ship terminal, Gold Coast, Australia 116, *129*, *132*

Kirra Groyne, Gold Coast, Australia 113–15, *132*

koa tree 49

Koholokai, K. 26–7

Kurungabaa 202

Kuta, Bali, Indonesia 157, 167–8, 184–5

Kyle, G. 206

La Herradura, Peru *129*

La Pampilla beach, Peru *129*

Labor Party, Australia 133

Laderman, Scott 12

laminating *see* glassing

Lanagan, D. 150

land tenure systems 126–7, 130

Langseth, T. 158

lapping 91

Lawler, K. 14, 44

Lawson-Remer, T. 153

Lazarow, Neil *11*, 28–9, 30–1, 32, 137

leashes 53

Leonard, T. 52

Leopold, Aldo 5, 170

leukaemia 94

lifesaving clubs 32

linseed oil 96

Lobitos, Peru 32

local communities: community development 33; cultural colonialism 53, 184–5, 186–91, **192–5**; displacement of 169; and surf parks 232–3; surf tourism 32, 33; traditional resource custodians 33–5

localism 47, 64, 147–62, 183, **184**, 246–7; challenges of 153–6; and colonialism 149–50, 157; and commercialisation 150; and crowding 150–2, 160–1, 165–6, 171; and cultural sustainability 152–3; different manifestations of 156–8; and environmental sustainability 153; in Nicaragua and Costa Rica 158–61; and surf etiquette 157; surf etiquette 148, 151–2, 154

Lochtefeld, Tom 220, 227, 228–9, 231

Long Beach, California *130*

Luddism 64–5

Luddites 47

Ludwig II of Bavaria 219

Luff, John 227, 230, 231, 232

Lugar de baixo, Madeira, Portugal *129*

McDonald, M. 225, 226–7

Machado, Rob 62

McKinnon, Andrew 114, 116

Macnaughten, P. 179

Madeira, Portugal *129*

Maldives *131*, 225

Malibu Beach, California 167–8

Mallacoota, Victoria, Australia *129*

Manning, R. E. 204–5

manufacturing *see* surfboard manufacturing

Marine Protected Areas (MPAs) 28, 174

Maritime Safety Queensland (MSQ)
112–13
Marko Foam 96, 99
Martin, Steven Andrew 13, 178
Martin's Beach, California *131*
Marx, Karl 45
Matuse 63
Mavericks, California 140, 223
maximum willingness-to-pay 139
meaning and culture 202–16, 248;
    attachment to place 205–6; diversity
    215; motivation for surfing 207–9, *207*,
    208, *209*, 214; serious leisure 205, 214;
    social and environmental perceptions
    209–13, **211**, **212**
media, surf 43–4, 148, 202, 226
Mentawai Islands, Indonesia 32, 35, 225,
    226
Mexico 59
Mill, J. S. 42
Miller, G. T. 31
Miller, M. L. 205
Minsberg, T. 59
modernity 4, 6–9, 13, 15
Mol, A. 192–3
Monk, Walter 56
Montauk, New York *129*
Morning of the Earth 202
motivation for surfing 207–9, *207*, **208**,
    209, 214
mountain sports 73
multiple streams theory of agenda setting
    109, 120
Mundaka, Spain 31, 137, 223
Murphy, M. 31, 137
myeloid leukaemia 94
myths, Nirvanic 226

Narrowneck artificial reef, Gold Coast,
    Australia 118, 119
National Surfing Reserves (NSRs) 115, 121
natural capital 30, 31
natural resources: conservation 25–6,
    27–9, 153, 170, 174; traditional resource
    custodians 33–5; waves as 30
Nazer, D. 156, 157
Nelsen, Chad 137, 140
neoprene 53, 81–2
Netherlands 222
New Zealand 60, *129*, 157
Newman, Campbell 116
Newquay, England 167–8
NGOs *see* non-profit/non-governmental
    organisations

Nias, Indonesia 189
Nicaragua 153, 158–61
Nirvanic myths 226
Nirvanification 225–31
Noble, D. 42, 45, 46
non-market values of surfing 137–42, 223,
    246
non-profit/non-governmental organisations
    14, 24, 32, 33, 62, 63, 66, 95–6, 115,
    120–1, 129–30, *130*, 137, 171, 173, 202
non-use values 138–9
Norway 158, 220
Nourbakhsh, T. A. 154
nuclear power 8
Nye, D. E. 45, 47

Oahu, Hawai'i 138, 167–8
O'Brien, Danny 14
occupational health and safety, surfboard
    manufacturing 51, 83, 94–5, 99–101
Ocean Dome, Miyazaki, Japan 220
offshore developments 24
Offshore Technology Corporation (OTC)
    220
Oil Pollution Control Act (1990), USA
    139
oil spills 139
olo board 49
Olympic Games 4, 224
one surfer per wave protocol 166
oppositional postmodernism 15
Orams, Mark 155
Orbach, M. K. 206
Oregon 156
organic epochs 48
Ouhilal, Y. 176, 179, 180
Our Common Future 5, 24
Outdoor Sports Valley 73
overcrowding *see* crowding

packaging materials 62, 63, 82
Padang Padang, Bali, Indonesia 183, **184**
paddle out protests 129, *129*, 130
paddleboards, stand-up 52–3, 172
Pallisades Amusement Park, New Jersey
    220
Palm Beach Shoreline Project, Gold Coast,
    Australia 118, 119, *132*
Papua New Guinea 34–5
Papua New Guinea Surfing Association
    34
Parkinson, Joel 117
participation levels 107–8; *see also*
    crowding

Patagonia 63
paulownia (*Paulownia tomentosa*) 97
Paumgarten, N. 54–5
Pavones, Costa Rica *130*, 158–61
Pearson, Kent 207, *207*, 208
peer-to-peer information sharing 58–9, 65–6
Pendleton, L. 139
Peru 32, *129*, 137
Peterson, Clay 96
phenol 100
Philippines 91
photovoltaic cells 233
Pichilemu, Chile 137
Pipeline, Hawai'i 26
place of the tube *see* consequences of surfing travel
plastic bags 82
Playa Encuentro, Dominican Republic *131*
Pleasure Point, California 139
Plymouth Sustainability 61
pneumatic caisson pools 221, 233, 234
policy entrepreneurs 109
polishing 91
politics of pragmatism 196–7
pollution: impact on surfers 26, 210–11, **212**, 213; surfboard manufacturing 51–2, 89, 93
polyester resins 91, 93, 99
Polynesia 41
polystyrene *see* expanded polystyrene (EPS); extruded polystyrene (XPS)
polyurethane 51–2, 89, 90, 91, 93, 94, 95, 99
Ponting, Jess 13, 33, 34–5, 44, 58, 180, 186
Popoyo, Nicaragua 158–61
Portugal 60, *129*, 222
postmodernism 4, 12, 15
pragmatic politics 196–7
Preston-Whyte, R. 206
Project Wave of Optimism 174
Prometheus 67
property taxes 141–2
protest movements 66, 107–8, 114–17, 125–34, *129*, *130*, *131*, *132*, 153, 205, 245–6
PU foams 51–2, 89, 90, 91, 93, 94, 95, 99
public awareness 32
public sphere protest 66, 107–8, 114–17, 125–34, *129*, *130*, *131*, *132*, 153, 205, 245–6
Puerto Rico 153
Punta de Lobo, Chile 66

Queensland: paulownia cultivation 97; *see also* Gold Coast, Australia
Quiksilver Pro France 75

Ranch, The, California 130, 148
Rastovich, Dave 62
real estate values 138, 140–2, 223
'record everything' mindset 54–5
recreational succession model 168–9
recycling/recycled materials 62, 63, 80, 81–2, 95, 96, 99
Red Bull Unleashed contest 232–3
reefs 34–5; artificial 24, 118, 119; *see also* artificial surfing reefs (ASRs)
reflexive modernity 8, 12, 15
regulation: surf management *109*; surfboard manufacturing 88–9, 93–4, 95, 100
relational sensibility 4, 10, *11*, 178, 193–5
research 10–14, *11*
reserves, surfing 26, 27–9, 115–17, 171
resins: environmental issues 93; environmentally friendly 63, 80, 95, 96, 101; glassing 91, 94–5, 99–101; health and safety issues 94–5, 99–101; polyester 91, 93, 99
resource stakeholders 23, 29–36, 242
respiratory illness 94, 95
Reunion Island 223
Riedel, William 'Stretch' 101
Rip Curl 92
risk society 6–9, *9*, 15
Ritchie, R. 25, 26
river mouths 24
Ron Jon Surf Park, Orlando, Florida 221
rubber 53
rules/surf etiquette 148, 151–2, 154, 157, 166
Russia 222

safety issues: boat users 112–13; crowding 150–2, 166, 167; shark attacks 214, 223; surf parks 230; surfboard manufacturing 51, 83, 94–5, 99–101
Saint-Simon, H. 48
San Diego State University 58, 61, 224
San Mateo Creek, California 24
San Sebastian, Spain 32
Santa Cruz, California 26, 140, 141
Save Our Southern Beaches Alliance 132
Save the Waves Coalition 32, 115, 117, 129, 137
Save Trestles movement 66
Scarfe, B. 10, 31

Schabet, Fidelis 219
Scorse, J. 140
Scott, P. 158
Scripps Institute of Oceanography 56
SDOs *see* surf development organisations
    (SDOs)
Seeney, Jeff 116
serious leisure 205, 214
services, surf tourism 30, 33
sewage 130, *130*
shaping 90, 91–2, 94, 98, 100–1
shark attacks 214, 223
sheet wave pools 220
Short, A. D. 27, 28, 29
Shridath, R. 242
SHY Technology 92
SIMA *see* EuroSIMA; Surf Industry
    Manufacturers Association (SIMA)
Slater, Kelly 61, 63, 101, 110, 219, 224
Snyder, G. 197
social media 54, 59, 126, 132, 172, 176, 219
social theory 14
socio-cultural issues 108; protests linked to
    128, 130
sociology *11*, 14
soft-top surfboards 52
solar energy 233
South Africa *130*, 167–8, 223
Spaargaren, G. 192–3
Spain 31, 32, *129*, 137, 223
specialisation 205
Spit cruise ship terminal, Gold Coast,
    Australia 110–11, *132*
sport industry clusters 72, *75*; *see also*
    Aquitaine surf industry cluster
sport management *11*, 14
'Spot X' 166, 171
Stafford, Bill *132*
stand-up paddleboards (SUPs) 52–3, 172
Stebbins, R. A. 205
Stewart, Michael 11, 95
stoke 4, 10, 11, *11*, 12, 178
Stradbroke Island, Australia 223
Stranger, Mark 4, 10, 12, 15
Stuckey, Jann 111–12
sugarbeet oil 95
Summerland Wavepool, Japan 220
Super Sap 95
SUPs *see* stand-up paddleboards (SUPs)
surf breaks *see* surf sites
Surf Cities 32, 33
surf development organisations (SDOs)
    33, 66, 173–4
surf etiquette 148, 151–2, 154, 157, 166

surf forecasting 56–7, 58, 63–4, 172, 176,
    225
surf imagery 43–4, 54–5, 58–9
surf industry clusters *see* Aquitaine surf
    industry cluster
Surf Industry Manufacturers Association
    (SIMA) 73, 77, 95
surf lifesaving clubs 32
Surf Lifesaving movement 125
surf localism *see* localism
surf management *11*, 107–9, *109*; *see also*
    Gold Coast Surf Management Plan
surf media 43–4, 148, 202, 226
Surf Park Central 224, 227
Surf Park Solutions 227
Surf Park Summit 224, 227
surf pools and parks 59–60, 66–7, 173,
    219–35, 249; history of 219–22; holy
    grail effect 224–5; key symbolic
    elements of 227–31; Nirvanification
    225–31; social and political factors
    222–4; surfers' concerns about 231–3;
    sustainable practices 233–4
surf porn 58
surf rage 148, 150, 153, 168, **212**, 213
Surf Resource Network 174
Surf Resource Sustainability Index (SRSI)
    26
surf schools 150, 154
surf sites: campaigns to protect 66, 107–8,
    114–17, 125–34, *129*, *130*, *131*, *132*,
    153, 205, 245–6; conservation 25–6,
    27–9, 153, 170, 174; demarcation of
    26–7, 149; exclusive access at 224–5;
    keeping information secret 47, 64, 166,
    171; management of *11*, 107–9, *109*;
    non-market values 137–42, 223, 246;
    physical boundaries 23–9, 241–2; real
    estate values 138, 140–2, 223; resource
    stakeholders 23, 29–36, 242; social
    dimensions 27; surfing reserves 26,
    27–9, 115–17, 171; urbanisation 13; *see*
    also crowding; localism
Surf Snowdonia, Conwy, Wales 221, 224,
    225, 227, 229–30, 232–3, 234
surf system boundaries 23; physical
    dimensions 23–9, 241–2; stakeholder
    dimensions 23, 29–36, 242
Surf Tech 220
surf tourism/travel *11*, 13, 222; bubble-
    style 34, 54, 58–9; carbon offsetting
    196–7; and conservation 25, 27; and
    crowding 151, 155, 169, 182–4;
    economic value of 137, 222–3;

employment 29; environmental impacts 53–4, 181–2, *183*, 185–6, **187–91**; and localism 155–6, 157, 159–60; and shark attacks 223; stakeholders 32–3; and surf forecasting 172, 176, 225; and surf imagery 58–9; surfing events 29–30, 80; and technology 53–4, 55, 65, 172–3; tourism development models 167–70, 247; traditional resource custodians 33–5; wilderness surfing experiences 165–7, 170–4; *see also* consequences of surfing travel
surf wax 81
surf zone 24, 27
Surf+SocialGood conference 61
SurfAid International 32, 33, 173–4
Surf-a-Torium, Japan 220
surfboard manufacturing 50, 51–3, 87–102, 243–4; ancient Hawai'i 48–9, 87; environmental issues 93–4; environmentally friendly materials 62, 63, 80–1, 95–7, 99, 101; glassing 91, 92, 94–5, 99–101; health and safety issues 51, 83, 94–5, 99–101; outsourcing 62–3, 80, 91–2; regulation 88–9, 93–4, 95, 100; shaping 90, 91–2, 94, 98, 100–1; subcultural origins and economic constraints 97–101; sustainable practices 62, 63, 80–1, 82–3, 95–7, 99
surfboards: foam/fibreglass 51–2, 87, 88–9, 90–2, 93; soft-top 52; stand-up paddleboards (SUPs) 52–3, 172; wooden 48–9, 63, 87, 96–7
Surfbreak Protection Society 171
SurfCredits 174
surfers: attachment to place 205–6; beginners 154–5, 156, 231; with disabilities 153–4; diversity 215; economic impact 29; ethics 32; motivation for surfing 207–9, *207*, **208**, *209*, 214; peer-to-peer information sharing 58–9, 65–6; 'record everything' mindset 54–5; as resource stakeholders 31–2; serious leisure 205, 214; social and environmental perceptions 209–13, **211**, **212**; US average 22; and waves 10–12, *11*; *see also* localism
Surfers Against Sewage 32, 129–30, *130*, 171, 173, 202
Surfers Environmental Alliance 32
Surfers Journal, The 202
Surfer's Paradise, Queensland, Australia 167–8
Surfer's Path, The 202

surfer-volunteerism programmes 33
surfing, defining 3–4
Surfing Australia 118
Surfing Capital 30–1, 108, 125; campaigns to protect 125–34, *129*, *130*, *131*, *132*; management strategies 108, *109*
surfing common 126–8, 130, *131*
surfing competitions *see* surfing events and contests
surfing events and contests 29–30, 80; ancient Hawai'i 49; broadcasting on Internet 59; and crowding 111–12; economic value 138; environmentally friendly boards 101; surf parks 220, 232–3
Surfing for Autism 153
surfing habitat 25; conservation 25–6, 27–9, 153, 170, 174
surfing industry *11*, 12, 14–15, 29, 150, 222; Gold Coast, Australia 118; outsourcing 62–3, 80, 91–2; services 30, 33; surf imagery use 43–4, 58–9; sustainable production 62–3, 80–1; *see also* Aquitaine surf industry cluster; surf pools and parks; surf tourism/travel; surfboard manufacturing
surfing participation levels 107–8; *see also* crowding
Surfing Queensland 118, 121
surfing research 9–14, *11*
Surfing Research Group 61
surfing reserves 26, 27–9, 115–17, 171
surfing utopia *see* consequences of surfing travel
Surfline 56, 63
Surfloch 221, 222, 228
surf-movie genre 165
surfonomics *see* economics of surfing
Surfrider Foundation 24, 32, 129, 140, 171, 173, 202
Surfrider Foundation Australia 120–1
surrounding gaze politics 66
sustainability certification programmes, surf tourism-specific 33
sustainability era, shift towards 43, 60–7
sustainability indicators 26
sustainable business practices 72–3; collective 81–3; firm-level 80–1
sustainable development 4–6; defined 5; and risk society 7, 8–9, *9*, 15
Sustainable Stoke Conference 58, 222
Sustainable Surf 25, 62, 63, 95–6
system boundaries *see* surf system boundaries

Tahiti 59
Tate, Tom 110–11, 113, 117, 118, 121
Tavarua Island Resort, Fiji 224–5
TCM *see* Travel Cost Method (TCM)
TDI *see* Toluene Di Isocyanine (TDI)
technique 45
technological determinism 45–7
technological momentum 47
technology 41–67, 242–3; artificial surfing
    51, 59–60, 66–7; climatology **51**, 56–7,
    63–4; ecologically connected imagery
    43–4; epistemological Luddism 64–5;
    image capturing 47, 54–5, 172; physical
    dimension 50–5, **51**, 62–3; shift towards
    sustainability era 43, 60–7; surf
    forecasting 56–7, 58, 63–4, 172, 176,
    225; and surf tourism/travel 53–4, 55,
    65, 172–3; sustainable life-cycle
    production 62–3; transportation 53–4,
    176–7, 196; *see also* equipment; Internet
    communication technologies (ICTs);
    surfboard manufacturing
technology, environment, and society
    (TES) framework 48–50
Telo Islands, Indonesia 225
Thailand 91, 92
Thanburudhoo, Maldives *131*
Thoman, D. 156, 158
Tilley, Charles 139
timber surfboards *see* wooden surfboards
Toluene Di Isocyanine (TDI) 89, 93, 100
Tomson, Shaun 223
Torquay, Victoria, Australia 78
tourism *see* surf tourism/travel
tourism destination life-cycle model
    167–8, 169
tourism development models 167–70, 247
Tourism Gold Coast 131
Towner, N. 35, 155
toxic chemicals 51–2, 89, 93, 94–5,
    99–101
Tracks (magazine) 202
traditional resource custodians 33–5
transpersonal ecology 195–6
transportation technology 53–4, 176–7,
    196
travel *see* surf tourism/travel
Travel Cost Method (TCM) 139–40
Trestles, California 24, *131*, 137, 140
Tuan, Y. F. 192
Typhoon Lagoon, Disney World Orlando,
    Florida 220, 221

UK *see* England; Wales

Uluwatu, Bali, Indonesia 26, 167–8, 180,
    181, 223
United Arab Emirates 221
United Nations 5, 8
United States: artificial surfing reefs 60;
    average surfer 22; campaigns *129*, *130*,
    131, 153; economic value of surfing
    137, 138, 139–40, 141, 222, 223;
    localism 149–50, 152, 155–6, 158; surf
    pools 220, 221, 222; surfboard
    manufacturing 88–9, 90, 95, 96, 97, 98,
    100, 101
United States Surfing Federation 153–4
University of California Berkeley 93
urbanisation 13
utopian spaces of surfing *see* consequences
    of surfing travel
UV-cured resin system 96

Valderrama, A. 153
Vanderburg, W. 44, 45, 46, 57, 58
Vans Triple Crown of Surfing 101, 138
Venus Grotto 219
Victoria, Australia 27, 78, *129*
violent incidents 148, 149, 150, 153, 156,
    212, 213
virtual-reality surfing experiences 173
volatile organic compounds (VOCs) 93, 96
volunteerism 33
Vorster, Herman 116

Wadi Adventure surf pool, United Arab
    Emirates 221
Waikiki, Hawai'i 26
Waitt, G. 149, 157
Wales, Surf Snowdonia 221, 224, 225,
    227, 229–30, 232–3, 234
Walker, I. H. 149
Walker Foam 94, 95
Ware, Dan 121
Warren, A. 12, 88–9, 149
Warshaw, M. 178, 204, 220
Washington 156
waste: image capturing devices 54;
    shipping out of island resorts 186, **188**,
    189; surfboard manufacturing 51–2, 53,
    93, 96
Waste to Waves programme 96
water drop surf pools 220, 221
water quality: campaigns to protect 128,
    130, *130*; impact on surfers 26, 210–11,
    212, 213
water theme parks *see* surf pools and parks
waterproof cameras 47, 54–5, 172

'Wave, The', Bristol, England 229
wave access: exclusive 224–5; *see also* crowding; localism
Wave Energy Converters (WECs) 24
wave forecasting 56–7, 58, 63–4, 172, 176, 225
wave frequency 108, 128; surf pools 220, 221, 230, 231
wave pools *see* surf pools and parks
wave quality 108, 128; and localism 156; surf pools 227–8, 231
Wavegarden 221–2
Waveloch 220, 228
waves: artificial 59–60; campaigns to protect 128, 129, *129*; as natural resource 30; sharing 166; and surfers 10–12, *11*
Waves for Development 32, 33, 174
Waves of Freedom 174
Webber, Greg 227–8, 231
Webber Wave Pools 221, 222, 227
Wegener, Tom 97
West, P. 178
Western appropriation era 41, 49–50
wetsuits 50, 53, 63, 81–2, 172

Whilden, Kevin 11, 95
Wildcoast 32
wilderness surfing experiences 165–7, 170–4, 247
wili wili tree 49
willingness-to-pay values 138, 139
Wilson, Jimmy 221
Winner, L. 64–5, 67
wooden surfboards 48–9, 63, 87, 96–7
Woody, T. 99, 101
World Commission on Environment and Development 45–6
World Professional Inland Surfing Championships 220
World Surf Cities Network 32
World Surf League (WSL) 59, 222, 223
World Surfing Reserves (WSRs) 115–17, 129
Wounded Warriors 154

XPS *see* extruded polystyrene (XPS)

Young, Nat 151
YouTube 59

Milton Keynes UK
Ingram Content Group UK Ltd.
UKHW040110071024
449327UK00019B/951